BIOCHEMISTRY AND MOLECULA
THE POST GENOMIC ERA

# GENE DELIVERY

# METHODS AND APPLICATIONS

# BIOCHEMISTRY AND MOLECULAR BIOLOGY IN THE POST GENOMIC ERA

Additional books and e-books in this series can be found on Nova's website under the Series tab.

BIOCHEMISTRY AND MOLECULAR BIOLOGY IN
THE POST GENOMIC ERA

# GENE DELIVERY

# METHODS AND APPLICATIONS

VANESSA ZIMMER
EDITOR

nova
science publishers
New York

Copyright © 2019 by Nova Science Publishers, Inc.

## NOTICE TO THE READER

## Library of Congress Cataloging-in-Publication Data


ISBN: 978-1-53616-268-4
Library of Congress Control Number:2019950413


*Published by Nova Science Publishers, Inc. † New York*

# CONTENTS

# PREFACE

Gene Delivery: Methods and Applications provides a comprehensive overview on viral and non-viral methods used to genetically engineer human mesenchymal stromal cells. In addition, an update on ongoing and completed clinical studies with engineered mesenchymal stromal cells will be provided, as well as a snapshot of the advances and technical challenges yet to be addressed.

Next, a variety of gene delivery systems including physical transfection techniques, virus-based delivery vectors, chemically engineered delivery systems and bio-inspired vehicles are reviewed and their strengths, shortcomings and biomedical applications are discussed.

Selfish DNA called transposons capable of cutting out and pasting into the host genome are active throughout the phylogenetic kingdoms. Researchers have repurposed natural transposons for use in delivering a gene-of-interest, enabling for the study of a large and growing list of preclinical gene therapy applications. As such, the authors discuss the past achievements and future challenges of this early-stage technology.

The closing chapter introduces cell-penetrating peptides as an efficient tool for DNA transfection. HR9, a designed cell-penetrating peptides, containing nona-arginine flanked by cysteine and penta-histidine displayed a high penetrating ability in mammalian cells.

Chapter 1 – Mesenchymal Stromal Cells (MSCs) are a diverse subset of adult multipotent precursors capable of differentiating into cell types of mesodermal lineage, such as adipocytes, chondrocytes and osteocytes. *In vivo*, MSCs are mostly located in perivascular locations, providing stromal support and maintaining a dynamic and homeostatic tissue microenvironment, mainly through the secretion of signaling molecules. MSCs can modulate several biological mechanisms through paracrine activity, namely limiting apoptosis, inducing angiogenesis and supporting the proliferation/differentiation of other stem/progenitor cells. In addition, the immunomodulatory potential as well as the site-specific migratory capability of MSCs have been demonstrated in several studies. In this context, MSCs show a great promise in the field of regenerative medicine, with potential for numerous clinical applications such as graft-versus-host disease (GvHD), and several cardiovascular, autoimmune and liver diseases. However, the heterogeneity of the MSCs populations administered into patients and the poor cell engraftment observed rendered some inconclusive results in clinical studies, thus limiting their use in cell-based therapies. Hence, genetic engineering appears as a promising strategy to overcome these limitations and maximize the therapeutic potential of MSCs. In the last decade, several studies employing MSCs as therapeutic vehicles of exogenous genes and drugs have been reported. Though viral systems have demonstrated the highest gene transfer efficiencies into human MSCs, non-viral vectors and gene transfer approaches are emerging as safer and effective alternatives.

This chapter provides a comprehensive overview on viral and non-viral methods used to genetically engineer human MSCs. In addition, an update on ongoing and completed clinical studies with engineered MSCs will be provided, as well as a snapshot of the advances and technical challenges yet to be addressed related to the development of MSC-based therapies.

Chapter 2 – Gene therapy is a major breakthrough in biotechnology nowadays. The premise for a successful gene therapy is the delivery of the therapeutic nucleic acids into target cells in a specific, effective and safe manner. With the development of biotechnology and nanotechnology, many delivery systems were developed. In this chapter, a variety of gene delivery systems (physical transfection techniques, virus-based delivery vectors, chemically engineered delivery systems and bio-inspired vehicles) will be reviewed and their strengths, shortcomings and biomedical applications will be discussed.

Chapter 3 – Selfish DNA called transposons capable of cutting out and pasting into the host genome are active throughout the phylogenetic kingdoms. Researchers have repurposed natural transposons for use in delivering a gene-of-interest. This has enabled the study of a large and growing list of preclinical gene therapy applications. Hyperactive forms of the transposase enzyme have been developed, enabling high efficiencies of integration in mammalian cells. Recently, transposons have entered clinical trials for delivery of CARs for T-cell based immunotherapy. Several groups have attempted to control where insertions occur in the genome. By tethering DNA binding proteins to the transposase, the engineered fusion protein can be directed to a desired sequence and caused to integrate the transposon nearby. A targetable transposase could overcome drawbacks with virus-based approaches, including limited cargo capacity, host immune response, and the risk of insertional mutagenesis. However, major hurdles will need to be overcome before targetable transposition becomes available to the clinic. The specificity of these first-generation vectors is poor, which has stimulated new research focused on reducing off-target integration. The long-term goal is to generate a vector that exclusively integrates at the target sequence and nowhere else. This review discusses the past achievements and future challenges of this early-stage technology.

Chapter 4 – Developing a useful and efficient DNA transfection method is always a concerning issue and necessary for study of specific molecules and their functions in individuals. However, most transfection methods utilized today are applied in mammalian cells such as embryonic cells and cell lines. Rare transfection studies are found in individual microscopic organisms such as paramecium and rotifer which belong to single-celled and multi-celled individuals, respectively. Here, we introduced cell-penetrating peptides (CPPs) as an efficient tool for DNA transfection. HR9, a designed CPP, containing nona-arginine flanked by cysteine and penta-histidine displayed a high penetrating ability in mammalian cells. Moreover, HR9 was able to internalize paramecia and rotifers which contain the pellicles and cuticles, respectively. DNAs were also delivered into these cells and organisms by HR9 and still contained the bioactivities. High viabilities of organisms and low cytotoxicities after HR9/DNA treatments illustrated that this CPP was harmless and could be a potent tool for DNA transfection.

In: Gene Delivery
Editor: Vanessa Zimmer

ISBN: 978-1-53616-268-4
© 2019 Nova Science Publishers, Inc.

*Chapter 1*

# GENE DELIVERY AS A TOOL TO IMPROVE THERAPEUTIC FEATURES OF MESENCHYMAL STROMAL CELLS: METHODS AND APPLICATIONS

*Marília Silva[1], Cristiana Ulpiano[1], Nuno Bernardes[1], Gabriel A. Monteiro[1,2] and Cláudia Lobato da Silva[1,2]\**
[1]Department of Bioengineering and iBB-Institute
for Bioengineering and Biosciences, Instituto Superior Técnico,
Universidade de Lisboa, Lisboa, Portugal
[2]The Discoveries Centre for Regenerative and Precision Medicine,
Lisbon Campus, Instituto Superior Técnico,
Universidade de Lisboa, Lisboa, Portugal

## ABSTRACT

Mesenchymal Stromal Cells (MSCs) are a diverse subset of adult multipotent precursors capable of differentiating into cell types of

---

\* Corresponding Author's Email: claudia_lobato@tecnico.ulisboa.pt.

mesodermal lineage, such as adipocytes, chondrocytes and osteocytes. *In vivo*, MSCs are mostly located in perivascular locations, providing stromal support and maintaining a dynamic and homeostatic tissue microenvironment, mainly through the secretion of signaling molecules. MSCs can modulate several biological mechanisms through paracrine activity, namely limiting apoptosis, inducing angiogenesis and supporting the proliferation/differentiation of other stem/progenitor cells. In addition, the immunomodulatory potential as well as the site-specific migratory capability of MSCs have been demonstrated in several studies. In this context, MSCs show a great promise in the field of regenerative medicine, with potential for numerous clinical applications such as graft-versus-host disease (GvHD), and several cardiovascular, autoimmune and liver diseases. However, the heterogeneity of the MSCs populations administered into patients and the poor cell engraftment observed rendered some inconclusive results in clinical studies, thus limiting their use in cell-based therapies. Hence, genetic engineering appears as a promising strategy to overcome these limitations and maximize the therapeutic potential of MSCs. In the last decade, several studies employing MSCs as therapeutic vehicles of exogenous genes and drugs have been reported. Though viral systems have demonstrated the highest gene transfer efficiencies into human MSCs, non-viral vectors and gene transfer approaches are emerging as safer and effective alternatives.

This chapter provides a comprehensive overview on viral and non-viral methods used to genetically engineer human MSCs. In addition, an update on ongoing and completed clinical studies with engineered MSCs will be provided, as well as a snapshot of the advances and technical challenges yet to be addressed related to the development of MSC-based therapies.

**Keywords**: mesenchymal stromal cells, *ex vivo* gene delivery, viral vectors, non-viral vectors, cell and gene therapies

# 1. MESENCHYMAL STROMAL CELLS (MSCs): A BRIEF INTRODUCTION

Mesenchymal stromal cells (MSCs) are multipotent stromal cells that can differentiate into a variety of lineages, including osteocytes, adipocytes and chondrocytes [1]. This differentiation capability, along with the release of trophic factors [2], immunomodulatory properties [3] anti-fibrotic [4]

and anti-apoptotic [5] features, sparked the interest of the scientific community in the last decades. Cells with these features could potentially hold a great promise for cell therapies and tissue engineering, by replacing damaged tissues of mesodermal origin, to promote regeneration, and to treat immune-mediated diseases. As such, the number of clinical trials using MSCs has been rising almost exponentially since 2004 [6], achieving a total of 1023 studies in 2019, of which 285 have been completed to date (Figure 1).

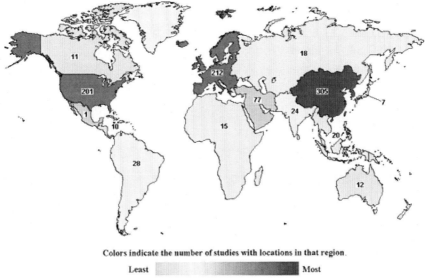

Colors indicate the number of studies with locations in that region.

Least ▭ Most

Labels give the exact number of studies.

Figure 1. Worldwide distribution of clinical trials obtained from "clinicaltrials.gov" accessed on June 17th, 2019, using the terms "mesenchymal stem cell OR mesenchymal stromal cell."

MSCs were described for the first time in 1970 by Friedenstein and co-workers, as colony forming fibroblasts (CFU-F), a rare population of cells residing in the bone marrow (BM) of guinea-pigs and mice [7, 8]. Although the term "mesenchymal stem cell" was suggested in 1991 by Arnold Caplan [9], the knowledge of the existence of MSCs was not new, pointing at the 19th century when the Germain biologist Cohnheim hypothesized that fibroblastic cells derived from BM were involved in wound healing throughout the body [10, 11]. Due to the increasing interest of researchers to study the therapeutic potential of MSCs, different methods of isolation and expansion, and different approaches to characterize the cells, led to difficulties in comparing the outcomes of different studies. To address this issue, in 2006, the Mesenchymal and Tissue Stem Cell Committee of the International Society for Cellular Therapy (ISCT) defined minimal criteria to characterize human MSCs. These criteria include: MSCs must be plastic-adherent when maintained under standard culture conditions; MSCs must express CD105, CD73 and CD90, and lack expression of CD45, CD34, CD14 or CD11b, CD79a or CD19 and HLA-DR surface molecules; and MSCs must differentiate into osteoblasts, adipocytes and chondroblasts *in vitro* [12].

Despite the first demonstration of the existence of MSCs in the adult BM, cells with similar characteristics have been found from cultures of virtually all adult and fetal organs tested [13]. In fact, MSCs can be found in nearly all tissues and are mostly located in perivascular niches. Some of the reported tissue sources include the BM [14], umbilical cord (UC) [15], Wharton' jelly (WJ) [16], placenta (PL) [17], menstrual blood (MB) [18], adipose tissue (AT) [19], gingiva (GV) [20] and dental pulp (DP) [21]. Since the pioneer studies by Friedenstein and colleagues [7], BM remains the gold standard as tissue source to obtain MSCs, although cells isolated from adipose tissue (AT-MSCs) and umbilical cord (UC-MSCs) have been also tested in cell therapy studies. At a first look, MSCs from different tissues share key characteristics such as fibroblast-like appearance *in vitro*, trilineage differentiation capacity (osteo-, adipo- and chondrocytic lineages), and expression of certain cell surface antigens [12]. However, some studies have stated that MSCs isolated from different sources in fact express different surface markers [22, 23], and may differ in what concerns differentiation potential. For instances, cells from these three sources present a similar capacity for chondrogenic and osteogenic differentiation, with the exception of UC-MSCs that show reduced adipogenic potential [24, 25]. Although BM has been the main source for the isolation of multipotent MSCs, the harvest of BM is a highly invasive procedure and the number, differentiation potential, and maximal life span of BM-MSCs decline with increasing age [25]. A significant advantage of the neonatal tissues, such as the UC, as sources of MSCs is that they are readily available, thus avoiding invasive procedures and ethical problems associated to adult tissues. It is also suggested that MSCs from these neonatal tissues may have advantages in comparison to MSCs derived from adult sources as BM. Indeed, several studies have reported superior proliferative capacity, life span and differentiation potential of MSC from birth-associated tissues over BM-MSC [25, 24]. AT is another alternative source to retrieve MSCs that can be obtained by a less invasive method and in larger quantities compared to BM. Cells form AT can be isolated from liposuctions in large numbers and grown easily under standard tissue culture conditions. Importantly, AT might be a robust cell source when envisaging an "off-the-shelf" product requiring mass cell production, due to the abundance, relatively easy harvest, and high MSCs frequency in this tissue [25]. Nevertheless, depending on the study focus and therapeutic target, MSCs from different tissue sources could serve different purposes and should be selected accordingly.

Despite the paracrine activity, by secreting a wide range of growth factors, cytokines, and hormones, and cell-to-cell interactions that play a key role in processes such as immunomodulation and tissue repair/regeneration, MSCs also rely on the secretion of extracellular vesicles (EVs) to communicate and exert their effects. Among EVs, exosomes have been extensively studied over the last years. These vesicles transport a variety of cellular components and mediators such as proteins, lipids, and many different species of RNAs to the neighbour or distant cells [26]. Interestingly, there are some studies suggesting that exosomes from MSCs might retain some of their mother-cell features, potentially able to exert a therapeutic effect [27–29] in a cell-free approach, which can obviate some of the problems associated with living products in a clinical scenario, namely safety.

# 2. MSCs AS A DELIVERY PLATFORM FOR CELL AND GENE THERAPIES

Beyond all the interesting features of MSCs previously stated, several studies have demonstrated the ability of MSCs to selectively migrate towards sites of injury [30]. This is one the most unique and attractive feature of MSCs in regard to cell therapies, commonly known as homing capacity or cell tropism. Stem cell homing is a phenomenon that was initially related to hematopoietic stem cells (HSCs) since these cells are able to migrate through the bloodstream to different organs and return to their niches at BM under the guidance of specific biochemical signalling [31]. Similarly to HSCs, MSCs also exhibit homing properties. This phenomenon allows BM stromal cells to migrate and engraft into damaged tissues throughout the body. When BM-MSCs are systemically administered after a stroke, the cells migrate and home towards the brain, improving the functional outcome [32]. Several animal studies indicate that direct delivery of MSCs to injured tissues can significantly promote their structural and functional recovery. These cells migrate and engraft into different tissues regardless the causes of the injuries and the tissue type [33]. While there is a lot still to uncover concerning the mechanisms underlying this remarkable and complex feature of MSCs, it is acknowledged that cells are specifically attracted to the sites of pro-inflammation and tissue damage, which is typically associated with cytokine signalling, and in this process the C-X-C chemokine receptor type 4 (CXCR4) - stromal cell-derived factor 1 (SDF1$\alpha$) axis has an important role [34]. *In vivo*, MSCs are able to secrete a wide variety of different growth factors, cytokines, and adhesion molecules by which they will affect the microenvironment of the inflamed and degenerating target tissue and thus maintaining a positive paracrine effect on the tissue repair [34].

The capacity of homing towards inflammatory sites along with the ability to modulate the defence mechanisms of the host, makes MSCs a promising therapeutic agent in maintaining transplantation tolerance, and exerting a therapeutic effect in multiple disease models, such as graft versus host disease (GvHD) [35], diabetes mellitus type 1 [36], experimental autoimmune encephalomyelitis (EAE) [37], contusive spinal cord injury and its subsequent inflammation-related damage [38], among others. Very importantly, MSC have been described as immune evasive, being able to escape, to a certain extent, the recognition mechanisms by the immune system due to low expression of MHC I and lack of expression of MHC class II along with other co-stimulatory molecules [39]. This reduced alloreactivity is a major benefit in terms of host compatibility issues, allowing MSC administration in a allogenic context [40]. In fact, to date, clinical trials have shown mild or no adverse effects from MSC treatment [41].

MSCs may also act as primary matrices in processes of tissue repair caused by inflammation and injury, such as bone regeneration [42] and cardiac injury repair [43]. Considering their homing capacity to tumour sites, a highly pro-inflammatory microenvironment, MSCs can also serve as the targeted carriers of therapeutic agents, as part of the tumour stroma in anticancer therapy [30].

Altogether, these features suggest MSCs as an ideal warhorse for cell and gene therapies and regenerative medicine settings. However, one of the major limitations that remains to be overcome is the low survival rate of MSCs after infusion. The viability of MSCs and their capacity to effectively home to lesions following administration into the recipient are limited, and the majority of the cells die within a few hours, even if the cells were delivered into the local lesion site [44]. Several studies have exploited genetic engineering approaches aiming at expanding the robustness and boosting the therapeutic profile of MSCs, and the field has progressively gained more significance over the last years. Essentially, by improving their innate secretory functions through cell modification, tissue repair can be accomplished in a more targeted manner, in the most diverse range of pathologies. This improvement can be achieved by overexpression or suppression of factors naturally secreted by MSCs, using expression vectors or RNA inference (RNAi) molecules, respectively. MSCs have also been studied as a cellular vehicle for gene-transfer applications either designed to replace a missing protein or used in a trojan horse way to express non-native therapeutic proteins, targeting and eliminating malignant cells. Although in preliminary stages, molecular editing of MSCs using clustered regularly interspaced short palindromic repeats (CRISPR) systems have also been exploited in order to improve their innate therapeutic properties (Figure 2).

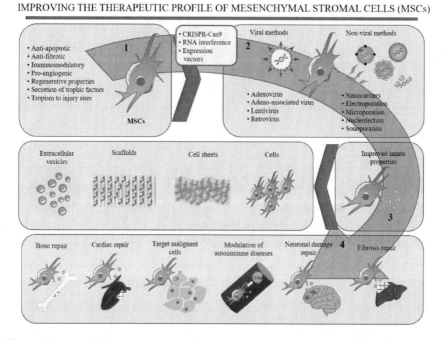

Figure 2. Improving the therapeutic profile of Mesenchymal Stromal Cells (MSCs) (1), through different genetic engineering approaches relying on viral and non-viral methods (2). Engineered cells can be delivered using different platforms (3), to treat several pathologies including ischemic diseases, tissue damage, autoimmune diseases and cancer (4).

In this chapter a comprehensive overview is provided on viral and non-viral methods used to genetically engineer MSCs and their potential applications will be provided. An update on the ongoing and completed clinical studies with engineered MSCs will be described, as well as a snapshot of the advances and technical challenges yet to be addressed related to the development of MSC-based therapies.

# 3. GENE DELIVERY METHODS FOR MSC-BASED THERAPY

## 3.1. Viral Methods

After the first ever approval in 2003 of an adenovirus-based gene product for head/neck carcinoma, Gendicine, in China [45] and with the first European marketing authorization in 2012 of an AAV1 vector for the treatment of a lipoprotein lipase deficiency (Glybera®) [46], the viral vector-based gene therapy is more and more rapidly evolving towards the first-line treatment of rare and acquired diseases for which different viral vectors systems are available. Being MSCs readily suitable to viral transduction, with optimized protocols that can prompt up to 90% transduced cells with no impact on multilineage differentiation capacity or the intrinsic properties [47], several are the studies that have exploited viral vectors to improve gene delivery. The use of viral vectors for gene transfer takes advantage of their natural ability to infect the cells. Therefore, transgenes may be incorporated either in addition to the genome or by replacing one or more genes. Viral transduction can thus offer a long-term and stable production of the protein of interest. These genetic modifications can then be targeted to improve the survival of MSCs, while mediating tissue repair and recovery in the *in vivo* environment, enhancing their therapeutic potential upon administration.

Currently, the most common viral vectors being used include retrovirus, lentivirus, adenovirus, and adeno-associated virus. Table 1 summarizes some of the main features of commonly used viral vectors in gene therapy, which are at the basis of the decision regarding the vector to employ according to the study's purpose.

**Table 1. Main characteristics of commonly used viral vectors for cell engineering [44, 45, 48]**

| Virus | Genome | Cargo capacity | Cells infected | Genome integration | Nature of Expression | Immune response | Main advantages | Main drawbacks |
|---|---|---|---|---|---|---|---|---|
| Adeno-virus | Single stranded DNA | > 8 kb | Dividing and non-dividing | No | Transient | Strong | Relatively large therapeutic genes Broad host range | Strong immune response |
| Adeno-associated virus | Double-stranded DNA | ≤ 4.5 kb | Dividing and non-dividing | 0.1% frequency | Stable | Very low | Several genome copies per target cell Generally safe at the doses tested | Immune response Complex production protocols Small transgenes |
| Retrovirus | RNA | < 8 kb | Dividing | Yes | Stable | Low | Efficient and stable gene transfer | Possible insertional mutagenesis |
| Lentivirus | RNA | < 8kb | Dividing and non-dividing | Yes | Stable | Low | Broad host range Low cytotoxicity High titers | Possible insertional mutagenesis |

In this section, a selection of studies employing genetic engineering to enhance the therapeutic efficacy of MSCs will be presented, focusing on the methods of viral modification. Overall, currently viral modified MSCs are playing a remarkable role in the treatment of a wide range of conditions (bone, cardiovascular, autoimmune, central nervous system diseases), including cancer. A summary of some of the applications of genetically modified MSCs using viral vectors is depicted in Table 2 (section 5.1.4).

### 3.1.1. Retroviral Vectors

Retroviruses comprise a large and diverse family of double-stranded RNA viruses having reverse transcriptase and a lipid envelope with receptor binding proteins. The hallmark of this family is its replicative strategy that includes as essential steps the reverse transcription of the virion RNA into linear double-stranded DNA and the subsequent integration of this DNA into the genome of the cell. After binding to the receptor, the external layer of the viral envelope fuses with the cellular membrane, internalizing the viral nucleic acid into the cytoplasm, which is then reversely transcribed into DNA that is integrated into the host genome [49]. Although this retroviral method results in a stable expression and long-term production of the desired protein, it could also lead to insertional mutagenesis and activation of oncogenes [50]. Also, retroviral vectors can only infect proliferating cells with very high efficiency. Post-mitotic cells such as myocytes or neurons are not susceptible to retroviral infection, thus limiting the spectrum of cells that can be targeted [49].

An example of a popular transgene incorporated into MSCs genome is tumour necrosis factor-related apoptosis-inducing ligand (TRAIL), known to induce apoptosis of tumour cells. Although TRAIL presents promising features, its clinical application is very limited due to the possibility of causing damage in normal tissues [51]. The intravenous administration of recombinant TRAIL contains several problems like short pharmacokinetic half-life and lack of targeted delivery, leading to its frequent and high dosage administration. The hypothesis of a cell delivery vector of this factor holds out to solve this constrains. In the study of Grisendi and colleagues, AT-MSCs were stably transduced with a retroviral vector. AT-

MSCs expressing TRAIL successfully targeted a variety of tumour cell lines *in vitro*, including human cervical carcinoma, pancreatic cancer, colon cancer, and, in combination with bortezomib, TRAIL-resistant breast cancer cells. Killing activity was associated with activation of caspase-8, as expected. When injected intravenously or subcutaneously into mice, TRAIL-MSCs homed into tumours and mediated apoptosis without significant apparent toxicity to normal tissues [52].

Also, several works with viral transduction include the incorporation of interleukins (ILs), a group of cytokines that mediate communication between cells. Although Interleukin 2 [33, 53] and Interleukin 6 [54] might promote tumour progression, there are multiple studies showing that MSCs genetically engineered to produce other ILs can successfully inhibit tumorigenesis. Although ILs are potent anti-cancer agents, when systemically administered at high dosages these became cytotoxic. Therefore, a specific and targeted administration of these proteins would be advantageous and MSCs can be promising to accomplish this task [55]. Interleukin-12 (IL-12) has demonstrated a potent anti-tumour activity in a variety of mouse tumour models, showing inhibition of local tumour growth. Envisioning an efficient IL-12 delivery system, Elzaouk and colleagues [56], stably transduced MSCs with a retroviral vector expressing murine IL-12, (IL-12)-MSC. The data indicate that treatment with MSCs expressing IL-12 led to a significant reduction of metastases in the lung in a prophylactic model where the MSCs were injected intraperitoneally. Interestingly, (IL-12)-MSC showed a clear anti-metastatic effect even when applied 18 days prior to tumour cell inoculation and led to a pronounced growth retardation of established subcutaneous melanoma, prolonging mice survival.

As previously mentioned, MSCs also produce a high amount of extracellular vesicles (EVs) that also display a therapeutic effect, being also a promising therapeutic approach. Since MSCs have the remarkable tendency to home into tumours, EVs, in particular exosomes, produced by MSCs may retain the homing properties of their parent cells. To test this hypothesis, Altanerova and colleagues (2018) [57] developed two prodrug suicide gene therapy systems: MSCs engineered to express fused yeast cytosine deaminase::uracil phosphoribosyl transferase (yCD::UPRT) with 5-fluorocytosine (5-FC) as a prodrug – yCD::UPRT-MSC/5FC system 9 and MSCs expressing thymidine kinase of the Herpes simplex virus with ganciclovir as a prodrug – tkHSV-MSC/GCV system. Both systems led to the expression of enzymes for conversion of a non-toxic prodrug to an efficient cytotoxic compound from a cell DNA integrated provirus. The authors demonstrated that MSCs with the yCD::UPRT gene integrated into the cell's DNA, under the control of a strong retrovirus promoter, express mRNA of the transgene and pack it into the cargo of released exosomes, and upon tumour cell internalization in the presence of prodrug, the suicide gene exosomes induced cell death by intracellular conversion of 5-FC to 5-FU. Importantly, these exosomes internalize into tumour cells delivering not only mRNA of the suicide gene, but other native mRNA and regulatory miRNAs as well, being able to induce and/or modify various molecular pathways in the cells [57].

Kalimuthu and colleagues showed promising results also taking advantage of viral transduction to incorporate suicides genes into MSCs [58]. Tetracycline-controlled transcriptional activation is a widely used method of inducible gene expression wherein transcription is reversibly turned ON or OFF in the presence of the antibiotic tetracycline or one of its derivatives (e.g., doxycycline; DOX). The first, and most commonly used, suicide gene is the herpes simplex virus-thymidine kinase (HSV1-TK), which converts the drug ganciclovir (GCV) into cytotoxic GCV-triphosphate. The combination of the suicide gene HSV1-TK with a tightly controlled system, such as Tet-On, which simply regulates by presence, or absence of DOX, could reduce the side effects associated with gene-based therapy. Suicide gene therapy also generates systemic toxicity and byproducts. Therefore, controlling suicide gene expression might be a valuable tool to prevent side effects. In this context, the aim of the study, was to develop therapeutic MSCs expressing an inducible HSV1-TK suicide gene, and to validate therapeutic gene expression *in vitro* using a mouse model of anaplastic thyroid cancer (ATC). The authors reported that the Tet-On suicide gene system in MSCs successfully exhibited an inhibition of cancer cell growth by a bystander effect after induction with DOX and with GCV treatment [58].

MSCs have been shown to reduce left ventricular (LV) remodelling and promote recovery of cardiac performance after myocardial infarction (MI) [59]. Intramyocardial injection has been one of the routes for administration of MSCs into the infarcted myocardium, allowing a localized delivery and enhanced retention of transplanted cells. However, this method is invasive and may require costly surgical intervention, can lead to arrhythmias, representing an extra risk to patients who already have a compromised cardiac function. On the other hand, systemic infusion of MSCs provides a non-invasive alternative that would be well tolerated by patients with advanced coronary disease or MI. Nevertheless, in this latter case, most of the infused cells have been found entrapped in the lungs and liver, and only a small fraction effectively home toward the ischemic myocardium [60]. This phenomenon may be explained by the decreased ability of the infused cells to respond to homing signals emanating from the ischemic myocardium [61], or by microcirculatory disruption upon administration due to MSCs large size when compared with coronary microvessels [62]. Aiming at the improvement of the intrinsic ability of MSCs to home into pro-inflammatory microenvironments, Cheng and colleagues [61] modified MSCs using retrovirus expressing CXCR4. *In vitro* migration assays showed that overexpression of CXCR4 enhanced the migration of MSCs toward SDF-1. Furthermore, intravenously infused MSCs overexpressing CXCR4 homed in significantly greater numbers toward the infarcted myocardium than unmodified MSCs did, leading to reduced LV remodelling and enhanced recovery of function [61].

### 3.1.2. Lentiviral Vectors

Lentiviruses are a retrovirus subfamily that can incorporate foreign DNA into cells and produce high titers of viral particles. These contain the reverse transcriptase enzyme that converts RNA into DNA before becoming integrated into the genome of the host, leading to stable transgene expression. Lentiviruses can transduce dividing and non-dividing cells, allowing them to infect populations such as stem cells, cardiomyocytes and other cells, without exhibiting an immune response following infection. Due to possible insertional mutagenesis events, lentiviral vectors have undergone through several modifications in order to increase biosafety and reduce pathogenic risks [63]. Non-integrating vectors and self-inactivating vectors, which lack the long terminal repeat sequences that regulate the expression of viral genes and integration into the host genome have been developed [63]. Due to their promising features, the use of lentiviral vectors has considerably increased as gene therapy vectors for engineering MSCs and other cells towards therapeutic applications. In the past, many preclinical and clinical trials have been conducted with retroviral vectors. However, although retroviruses are still used successfully, recently there has been a transition from retroviral vectors to lentiviral vectors. This tendency may be explained by the ability of lentiviruses to transduce nondividing cells because as their RNA can translocate across the nuclear envelope [64], their integration patterns are different from retroviral vectors and seem to be less risky with respect to insertional mutagenesis [65]. Finally, these vectors can be produced at high vector titer [66, 67].

An example of a transgene incorporated into MSCs through lentiviral transduction is interferon-β (IFN-β), which is a cytokine secreted by fibroblasts involved in a variety of biological processes but is mainly recognized as a potent immunomodulatory protein. IFN-β has been extensively studied in a variety of cancer studies for its antiproliferative and proapoptotic activity. However, injection of IFN-β alone for cancer is often accompanied by side effects. Taking advantage of MSCs characteristics, Chen and colleagues [68] engineered human UC-MSCs as cellular carriers of the IFN-β gene towards a potential treatment for non-small cell lung cancer (NSCLC). The IFN-β gene was transduced into UC-MSCs ((IFN-β)-MSCs) and subsequently injected into A549 lung cancer-bearing mice via the tail vein. The results showed that (IFN-β)-MSCs could reach lung cancer cells directly, inhibit their growth, promote their apoptosis, while avoiding the damage to organs produced by the administration of IFN-β alone [68]. Ahn and colleagues [69] also developed a protocol for producing canine AT-MSCs expressing β. In this study the authors showed the possibility of using a combination of cell-based therapy and chemotherapy in a canine malignant melanoma. AT-MSCs genetically engineered to produce INF-β migrated to the tumour site, reducing significantly tumour burden in an mice model [69]. This was one of the first studies demonstrating the efficacy of combining a systemic chemotherapy with a stem-cell based targeted delivery of a cytokine to a malignant canine melanoma in mouse. The same experimental setup was applied by Dembinski to successfully eradicate ovarian cancer [70].

Zhu and colleagues [71] used UC-MSCs as carriers for LIGHT, a cytokine included in the tumour necrosis factor (TNF) superfamily, that leads to the proliferation of T and B-cells, natural killer cells, monocytes, and granulocytes, and inhibits tumour growth by inducing apoptosis. Aiming to target gastric cancer, human MSCs were engineered to produce LIGHT in order to overcome the toxicity associated with its systemic administration and to allow its local secretion. The results demonstrated that these engineered cells could successfully home to the gastric tumour site due to the interaction between SDF-1α produced by MSCs and CXCR4, a receptor expressed by the tumour cells [71].

TNF-α is a cytokine secreted by macrophages that induces death of certain tumour cell lines, such as gastric cancer, by causing vasculotoxicity to the tumour cells. This property turns TNF-α a promising tool for anti-cancer therapies. However, as in many other anti-cancer agents already mentioned, its systemic administration would cause numerous adverse effects to the patient. In the study performed by Mao and colleagues [72], the manipulation of UC-MSCs towards the production of TNF-α to target gastric tumour cells, when subcutaneously injected into tumour models, was performed. The authors observed the tumour homing capacity of these cells and a significant tumour volume decrease, a strong suppressive effect on the tumour growth and increased tumour necrosis [72].

Type 2 diabetes mellitus (T2D) is a complex metabolic disease characterized by chronic hyperglycemia, insulin resistance, and pancreatic β cell destruction. Apelin, a newly identified adipokine produced and secreted by adipocytes, regulates insulin sensitivity and glucose metabolism [73]. An injection of apelin could restore hyperinsulinemia and insulin resistance, however, apelin has a short half-life (< 5min in humans), being rapidly cleared from the circulation and thus limiting its therapeutic benefit [74]. Envisioning a stable and robust expression of apelin, Gao and colleagues [75] engineered human Wharton' jelly MSCs (WJ-MSCs) to overexpress apelin through a lentiviral vector and transfused these cells into T2D rats. The results demonstrated a significant improvement in insulin sensitivity and glucose availability, and promoted endogenous pancreatic β cell proliferation, as well as a raise in the plasma content of insulin and C-peptide. Remarkably, these effects persisted *in vivo* up to 42 days after a single infusion of apelin-transduced WJ-MSCs [75].

Despite the robust regenerative potential of bone tissue, in the case of biological environments with inadequate blood supply and poor bone stock or fractures with mechanical instability, healing can be compromised leading to delayed union or non-union of the fractures. Inadequate bone repair is also seen in settings such as revision total joint replacement and spinal fusion. Autologous bone graft remains the gold standard for the treatment of such cases, although several drawbacks are still associated to this procedure. Bone morphogenetic protein 2 (rhBMP-2) is an Food and Drug Administration (FDA) approved osteoinductive agent for clinical use for anterior lumbar spinal fusion and treatment of open tibia fractures. However, it is associated with inconsistent clinical results and several complications due to the supraphysiologic doses needed to induce an adequate biologic response and the rapid release of the protein from the collagen carrier. In the study by Bougioukli and co-workers [76], the authors have addressed the aforementioned issue by engineering human UC-MSCs through transduction with a lentiviral vector expressing BMP-2, leading to sustained production of an osteoinductive signal that could induce robust bone formation *in vivo* . The results demonstrated that these cells produced a sustained, long-term release of the osteoinductive cues, inducing heterotopic bone formation in a muscle pouch mouse model, providing a proof of concept for potential clinical applications of BMP-2-transduced UC-MSCs. A similar approach was pursued by Kim and colleagues [77] where canine MSCs were engineered to overexpress bone morphogenetic protein 7 (BMP-7) by a lentiviral-mediated gene transfer into AT-MSCs. These cells were cultured into cell sheets (BMP-7-CSs) and transplanted into critical-sized bone defects in dogs. BMP-7-CS combined with demineralized bone matrix (DBM) stimulated new bone and blood vessel formation, providing a strong osteogenic and vascular potential. Also, aiming at solving the burden of avascular necrosis of the femoral head (ANFH), Zhang and collaborators [78] studied the therapeutic effect of combining rabbit BM-MSCs expressing basic fibroblast growth factor (bFGF) with xenogeneic antigen-extracted cancellous bone (XACB) in bone tissue engineering. The results were very promising regarding improvement of angiogenesis and osteogenesis in ischemic necrosis areas, improving significantly the repair effect in an ANFH rabbit model.

Among cardiovascular diseases, stroke is one of the leading causes of death and disability worldwide. Several studies evidenced that MSCs administration significantly improves functional recovery after stroke in rats, as well as in the context of neurodegenerative diseases as Alzheimer's, although lower survival of MSCs limits their therapeutic effect. Signalling of extracellular regulating kinases 1/2 (ERK1/2) is crucial for cell survival, differentiation, and proliferation. Therefore, one strategy would be to modulate the intrinsic properties of MSCs by overexpressing these genes. This rationale was implemented by Gao and colleagues [75] to increase the viability of these cells after infusion, exerting more efficiently their role in rats post stroke. The results showed that cultured (ERK1/2)-MSCs displayed better viability, and more cells were observed after administration. Notably, the treatment with ERK-MSCs demonstrated higher efficiency in behavioural recovery at 7 and 14 days, when compared with to non-modified MSCs. The underlying mechanisms of this therapeutic benefit may be multifaceted including enhanced endogenous cell proliferation, neuroinflammation modulation, among other processes [75].

As mentioned in the previous section, TRAIL is a powerful factor known to induce apoptosis of tumour cells in diverse tumour models. Several are the studies employing MSCs as living carriers of TRAIL directly in the tumour microenvironment, resorting to lentiviral vectors.

Loebinger and colleagues [79] suggested MSCs as a delivery vehicle for this factor given their successful features in the field. They developed human MSCs engineered to produce and deliver TRAIL (TRAIL-MSCs) to a lung metastatic cancer model. These cells were able to kill cancer cells *in vitro* via the extrinsic apoptotic pathway to a higher degree than the recombinant protein control and they reduced the growth of early subcutaneous tumours and metastasis in *in vivo* mice models. This was the first study in the literature demonstrating a substantial reduction in metastatic tumour burden with frequent eradication of metastasis using MSCs expressing TRAIL and shows the promising potential of genetically engineered MSCs therapeutic applications for the treatment of primary tumours and their metastasis [79]. From the study of Loebinger and colleagues [79], a clinical trial was initiated in 2017 (see section 6). Substantial work has been developed in this topic with the same promising outcomes, including for hepatocellular cancer [80, 81], pancreatic cancer [82], pancreatic cancer with a combinatory approach of TRAIL-MSCs with paclitaxel (PTX)-based chemotherapy [83], tongue squamous cell carcinoma [84], colorectal carcinoma [85] and Non-Hodgkin's Lymphoma [86].

It has been also demonstrated that MSCs can successfully carry miRNAs (miRs) to modulate gene expression. miRNAs refer to a group of conserved, short non-coding RNAs of 20-22 nucleotides that are closely associated to regulate gene expression at the post-transcriptional level. miRNAs modulate numerous cellular activities by binding to the 3'-untranslated region (3'-UTR) of their target mRNAs. Therefore, an interesting application would be to engineer MSCs to overexpress a given miRNAs to regulate the expression of target genes that serve as potential therapeutic targets. In the study of Liu and co-workers [87], aiming at the improvement of bone regeneration in bone defects of calvaria in mice, mice BM-MSCs were transduced with miR-26a. This particular miR is involved in osteogenic differentiation by targeting SMAD family member 1, as miR-26a is upregulated during osteoblast differentiation. In this study, β-tricalcium phosphate (β-TCP) scaffolds loaded with miR-26a-modified BM-MSCs were implanted into cranial bone defect areas of mice. The regeneration and volume of newly formed bones were markedly increased when compared to the control groups, leading to an increase in protein expression of osteogenesis-associated cytokines. Other approach is to deliver miRs in exosomes produced by MSCs. In the study of Lang and colleagues [88], MSCs were transduced with lentiviral vectors containing miR-124a, an effective anti-glioma agent, and isolated vesicles (Exo-miR124) from the medium. Exo-miR124 exosomes were able to inhibit the growth and clonogenicity of glioma stem cells (GSC) and, when administered systemically, were capable of treating mice harbouring intracranial GSC xenografts. This was a pioneer study demonstrating that miR-124a–carrying exosomes inhibits GSC survival in both *in vitro* assays and *in vivo* animal models.

In order to support a robust tissue engineering technology in larger and more complicated tissue defect repairs, an extensive vasculature is needed to provide cells with the required nutrients. In the specific case of Periodontitis, one of the most common chronic oral diseases, achieving effective bone regeneration and revascularization has become a major focus in the field. Hypoxia-inducible factors (HIFs) play a crucial role in revascularization activating the expression of several downstream angiogenic factors, being HIF-1α a major transcription factor in controlling the expression of several HIFs. Chen and colleagues [89] down-regulated the expression of the prolyl hydroxylase domain-containing protein 2 (PHD2), the main catabolic enzyme for HIF-1α, in rat BM-MSC , employing a lentiviral vector-mediated short-hairpin RNA, aiming to promote the accumulation of HIF-1α and to enhance the angiogenic effects of these cells. A tissue engineered construct composed by a collagen membrane and the engineered MSCs was implanted in a rat periodontal fenestration defect model. The results of this study demonstrated that the engineered MSC had enhanced osteogenic differentiation in vitro and the resistance of cells to oxidative stress was also validated in vitro. Moreover, these cells demonstrated an enhanced promotion of neovascularization and repair of the periodontal fenestration defect in vivo [89].

So far, lentiviral vectors have been one of the most employed viral vectors in what concerns cell therapies with MSCs, mainly due to their prolonged, stable, and strong transgene expression. Nevertheless, there are some reports suggesting that there is some adverse impact of lentiviral transduction on the properties of MSCs [90]. Aiming at the optimization of the lentiviral transduction parameters of human endometrium-derived MSCs, Deryabin and colleagues [91] developed a new strategy and applied CRISPR-Cas9 technology for genome and secretome modifications as a model to verify the effectiveness of the procedure. Decidualization of endometrium is known to be an essential process for embryo implantation, placenta forming, and maintenance of pregnancy. In this context, a tightly controlled plasminogen activator inhibitor-1 (PAI-1) expression level by decidual cells has been reported to play a crucial role in this process. Therefore, MSCs could be applied for cell therapy of infertility associated with decidualization insufficiency. As such, in this study the authors selected PAI-1 as an engineering target, by developing a PAI-1 overexpressing human MSC line and a PAI-1 knockout MSC line through CRISPR-Cas9, since a rhythmicity of PAI-1 expression and secretion levels occurs in intervals throughout the pregnancy. To increase the lentiviral infection efficiency, the authors tested two additives: polybrene (Pb), also known as hexadimethrine bromide, and polycationic peptide protamine sulfate (Ps), commonly used to enhance transduction efficiency [90]. MSCs infected in the presence of Pb exhibited reduced proliferation and migration rates, increased senescence, and impaired adipo- and osteogenic differentiation ability, although with no impact in what concerns decidual differentiation. On the other hand, transduction with Ps had little impact on the identity criteria of MSCs, including surface marker expression profile, differentiation, migration and proliferation capabilities while maintaining higher infection efficiency. With this system, the authors were able to obtain both cell lines successfully, representing the first evidence of the effective MSCs secretome engineering via CRISPR-Cas9 genome editing technology [91].

### 3.1.3. Adenoviral Vectors

Adenoviral vectors have been one of the most explored viral methods in gene therapy settings. These are medium-sized (90–100 nm), nonenveloped (naked) viruses composed of an icosahedral protein capsid that encompasses a linear double-stranded DNA genome of approximately 36,000 base pairs. These vectors can be produced at high titers and hold a large amount of foreign DNA with the ability to infect both postmitotic and dividing cells. When these viruses infect a host cell, the viral DNA is introduced into the host without integration in the genome. The DNA molecule is left free in the nucleus of the host cell, leading to a transient expression. The transgenes are not replicated during cell division, therefore the genetic material will not be passed to the progeny cells. As a result of this, treatment with the adenoviral vectors will require repeated administrations of the therapeutic product. Adenoviral vectors are strongly immunogenic and induce a strong innate immune response, being this one of the main drawbacks for their therapeutic application in clinical practice [92]. However, improvements have been made to reduce the observed limitations of first-generation adenoviral vectors. A new class of adenoviral vectors is known as "gutted," "gutless," "helper-dependent" or "high capacity" adenovirus. In this gutted virus, all viral coding sequences are removed leading to a greater carrying capacity for foreign DNA, reduced toxicity, increased cloning capacity and prolonged transgene expression, making them suitable for clinical translation [93, 94]. However, these gutless adenoviral systems are difficult to produce in large quantities [95].

MSCs can be transduced by adenoviruses *in vitro*, but high viral titers are necessary to achieve high efficiency. However, according to Marasini and colleagues [96], transduction at a higher multiplicity of infection (MOI) is associated with attenuated proliferation and senescence-like morphology of transduced cells, as well as a diminished capacity for adipogenic differentiation compared to other lineages.

Adenoviral vectors are the most commonly employed vector for cancer viral therapy [92]. In this context, Kanehira and co-workers [97] developed mice BM-MSCs producing Interleukin-32, an antagonist of hepatocyte growth factor (HGF), which is a strong inducer of tumour growth, angiogenesis and lymphangiogenesis. Interleukin-32 also has inhibitory effects in angiogenesis triggered by other factors such as bFGF and vascular endothelial growth factor (VEGF). The results showed that these engineered MSCs could successfully migrate to the tumour sites inhibiting tumour progression in the lung and prolonging the survival of the animals tested. Also, in the context of cancer therapy, osteoprotegerin (OPG) is shown to inhibit the progression of osteosarcoma. MSCs transfected with adenoviruses carrying the OPG gene were injected into the tail vein of mice bearing osteosarcoma xenograft tumours. Administration of engineered MSCs was shown to reduce both tumour growth and to help limit general bone destruction [98]. In a similar approach, mice BM-MSCs engineered to express pigment epithelium-derived factor (PEDF) were tested in a model of Lewis lung carcinoma. Systemic application of PEDF-expressing MSC reduced tumour growth and prolonged survival of the experimental animals. Immunohistochemistry analysis showed enhanced apoptosis and decreased micro vessel density in the treated tumour-bearing animals [99].

Viral-based therapy for cancer is an old concept that has been revisited in recent years using genetically modified viruses, including oncolytic adenoviruses (OAd), meaning that the replication of these viruses can be restricted to malignant cells. At the end of the viral replication cycle, the tumour cell is lysed, and new infectious viral particles are released, which then infect and lyse neighbouring tumour cells. Castleton and colleagues [100] provided evidence that human MSCs can be efficiently used as carriers for systemic delivery of oncolytic measles virus to treat acute lymphoblastic leukemia, even in the presence of high-titer neutralizing antibodies. In a very interesting approach, Kaczorowski and co-workers [101] employed human MSCs as carriers of improved oncolytic adenoviruses, in which the pro-apoptotic early adenoviral gene E1B19K was deleted or a TRAIL transgene was inserted (OAd Ad5/3-TRAIL). The delivery of the OAd Ad5/3-TRAIL construct by MSCs induced a strong anti-tumour response in a pancreatic ductal adenocarcinoma model. This therapeutic approach was also effective in tumourigenic cells with cancer stem cell (CSC) characteristics, which do not respond to conventional chemotherapy [101, 102].

Although osteoarthritis (OA) is typically characterized by loss or damage of articular cartilage, inflammation of the synovial membrane is a prevalent feature believed to contribute to both symptoms and disease progression. Due to their chondrogenic differential potential, MSCs have been proposed for the repair of damaged cartilage in OA. Also, MSCs can polarise pro-inflammatory macrophages towards an anti-inflammatory phenotype and suppress T-cell proliferation and have been specifically reported to exert anti-inflammatory effects on human osteoarthritic synovium *in vitro*. Following this rationale, Farrell and collaborators [103] engineered human MSCs to overexpress IL-10, a cytokine that suppresses pro-inflammatory cytokines production, and has been reported to significantly decrease the frequency of arthritis, delay the onset of the disease and reduce the severity of arthritic symptoms. IL-10 was also reported to induce proliferation and promote chondrogenic or hypertrophic differentiation of primary chondrocytes, being a promising therapeutic factor to treat arthritis. The injection of IL-10-overexpressing MSCs ((IL-10)-MSCs) into a murine collagenase-induced OA model, led to long-term reductions in the numbers of CD4+ and CD8+ T cells in the draining popliteal lymph nodes of the knees and activated CD4+ and CD8+ T cells in the inguinal lymph nodes. While there was no clear reduction in disease onset or progression, this approach could be used in combination with reparative approaches to result in a long-term reduction in joint inflammation. Following a similar strategy, Tian and colleagues [104] designed rat BM-MSCs towards the production of IL-10 to be tested in a rat rheumatoid arthritis model. In this study, the results were also promising regarding the decreased expression of inflammatory factors, and a higher repair rate of osteoarticular cartilage and a higher inhibition of synovial proliferation was observed in the group treated with (IL-10)-MSCs.

Therapeutic angiogenesis has been indicated to be a promising line of treatment for patients with ischemic diseases. Blood vessel formation not only requires endothelial cells, but also pericytes and smooth muscle cells. The synergistic pro-angiogenic effect of basic FGF (FGF2 or bFGF) and platelet-derived growth factor (PDGF-BB) in the revascularization process have previously been reported, and when simultaneously administered into ischemic tissues, the two factors can promote mature and stable vessel formation. By exploiting the intrinsic angiogenic potential of MSCs, Yin and colleagues [105] transduced human placental-derived MSCs with adenoviral bicistronic vectors carrying the FGF2 and PDGF-BB genes. These cells were intramuscularly injected into the ischemic limbs of New Zealand White rabbits, and four weeks after cell administration, angiographic analysis revealed significantly increased collateral vessel formation in the group injected with engineered MSCs compared to non-modified cells. Histological examination revealed markedly increased capillary and arteriole density, and the xenografted therapeutic MSCs survived for at least 4 weeks subsequent to cell therapy [105].

An effective treatment is still lacking for diabetic peripheral neurovascular disease (DPNV), one of the most common chronic complication in diabetic patients. Administration of autologous/allogeneic AT-MSCs can represent a promising strategy for DPNV since the disease is frequently correlated to obesity, being AT a convenient MSC source. However, some reports state that the infused AT-MSCs demonstrate unsatisfying viability, migration, adhesion, and differentiation *in vivo*, which reduce the treatment efficiency [106]–[109]. Netrin-1 is an axon-guiding factor that not only promotes neuronal migration and secretion in the central nervous system, but also regulates the survival, adhesion, migration, proliferation, and differentiation of the endothelial cells and stem cells in non-nerve tissue and inhibits their apoptosis. In a study by Zhang and colleagues [110], mice AT-MSCs were transduced with the Netrin-1 (NTN-1) gene ((NTN-1)-MSCs). Netrin-1 improved the proliferation, migration, adhesion, and angiogenesis of AT-MSCs and prevented their apoptosis induced by high glucose *in vitro*. Furthermore, (NTN-1)-MSCs were implanted into type 2 diabetic mice with sciatic denervation. The group receiving engineered MSCs demonstrated a significantly higher blood flow (accessed non-invasively by the laser Doppler perfusion index) and microvessel density of the hindlimb muscles, when compared with the non-modified MSCs. The histological assay demonstrated that (NTN-1)-MSCs cells survived in the chronic ischemic muscles and supported the formation of small capillaries.

Radiation therapy is a conventional approach used to treat tumours, including thoracic tumours [111]. Despite being effective, it is often accompanied by severe side effects on lungs named radiation-induced lung injuries (RILI), such as lung fibrosis due to increased collagen-1, collagen-3 and TGF-β. As previously mentioned in this chapter, one of the demonstrated features of MSCs is their anti-fibrotic capacity. However, the translation of this potential to treat RILI still needs further investigation. Decorin (DCN) is a natural inhibitor of transforming growth factor-β (TGF-β) signalling, where it binds to collagen and reduces the formation of collagen fibrils thus disrupting their organization [112]. Therefore, a potential approach to minimize the burden of radiotherapy side effects in patients with lung cancer is the administration of DCN-modified MSCs. This strategy was developed by Liu and colleagues [112] by modifying human UC-MSCs with replication-deficiency adenovirus expressing DCN. MSCs were intravenously administered into an experimental mouse model of RILI. Transplanted MSCs homed to injured lung tissues and produced the DCN protein. Upon infusion, both unmodified MSCs and DCN-expressing MSCs (DCN-MSCs) could alleviate histopathological injuries by reducing lymphocyte infiltration, decreasing apoptosis, increasing proliferation of epithelial cells, and inhibiting fibrosis at the later phase. However, treatment with DCN-MSCs resulted in much more impressive therapeutic effects. Also, treatment with DCN-MSCs, was very effective in attenuating inflammation by reducing the expression of chemokines and inflammatory cytokines and increased the expression of anti-inflammatory cytokines [112]. Also in the context of fibrosis, Klotho, a protein primarily expressed by renal tubular epithelial cells (TECs) that has antiaging properties, was shown to reduce renal fibrosis after acute kidney injury (AKI) and to inhibit stem cell senescence [113]. In this context, Zhang and co-workers [113] hypothesized that mice BM-MSCs modified to express the Klotho gene (Klotho-MSCs) would have the ability to promote injury restoration by an immuno-regulation effect. Upon infusion of Klotho-MSCs into mice with AKI, the engineered cells ameliorated the loss of renal function and renal interstitial fibrosis, and the renal protective effect of Klotho-modified BM-MSCs was more powerful than that of non-modified BM-MSCs. One of the potential limitations of an MSC-based therapy relies on cell propensity towards aging that leads to a decline in proliferative ability and function, compromising their therapeutic effect. One interesting observation in this study was that the proliferative ability of Klotho-MSCs was higher than non-engineered MSCs. The authors hypothesized that a potential mechanism by which Klotho expression is able to maintain the proliferative ability of BM-MSCs is correlated with the expression of pluripotency genes, as they observed increased mRNA levels of Oct4 and Nanog in MSCs expressing Klotho. Also, the engineered MSCs inhibited macrophage infiltration and downregulated Wnt/β-catenin pathway activation in TECs [114].

Articular cartilage, an avascular and aneural tissue, plays a physiological role in lubrication and cushioning during movement. However, as it lacks vascularization, articular cartilage has a limited capacity for spontaneous repair after injury. Therefore, cartilage tissue engineering is considered to be more suitable for articular cartilage repair. MSCs are commonly used as seeding cells in cartilage tissue engineering approaches. In comparison with chondrocytes, MSCs represent a promising cell source that may be conveniently manipulated to differentiate into chondrocytes and it has shown benefits in the treatment of focal cartilage defects and osteoarthritic lesions [115–117]. However, there is no evidence showing that the function of the original hyaline cartilage is completely restored in the treated patients, rather than a repairing process leading to the formation of a fibrotic tissue that cannot withstand mechanical loads over time [118]. C-type natriuretic peptide (CNP) plays an important role in the regulation of cartilage growth and can promote the proliferation of primary chondrocytes and chondrocytes derived from MSCs. In the study by Yang and colleagues [119], an innovative approach to target articular cartilage defects was developed by combining gene therapy with tissue engineering strategies to test their effect on cartilage repair. Rat MSCs expressing CNP were loaded onto a chitosan/silk fibroin (CS/SF) porous scaffold to test their effect on repairing full-thickness cartilage defects in rat joints. The results of gross morphology and histology examination showed that the composite of the CNP gene-modified BM-MSCs and CS/SF scaffolds had a better tissue repair performance and more cartilage matrix than the non-modified MSCs at each time point [119].

### 3.1.4. Adeno-Associated Viral Vectors

Finally, adeno-associated viral vectors are one of the most promising vectors as they are non-pathogenic to humans, have demonstrated safety in several clinical trials and results in long-term gene expression. The adeno-associated virus (AAV) genome is linear, consisting of a single stranded DNA molecule of approximately 4.7 kb, which contains two open reading frames (ORFs) flanked by 145 bp inverted terminal repeats (ITRs). AAV requires adenovirus or herpes virus as helper virus to initiate and complete its viral life cycle of rescue, replication, and packaging. Recombinant AAV-based vectors have the viral genome deleted, other than the ITRs, allowing for insertion of vector constructs to a maximum of approximately 4.9 kb. In their recombinant form, most of the AAV vectors persist as concatemers, forming linear and circular forms, which leads to long-term persistence of the viral genome in transduced cells. Of importance, AAV vectors can target postmitotic cells such as neurons [95]. However, a large fraction of the human population have neutralizing antibodies against AAV, which drastically reduces their *in vivo* efficacy [120]. Also, these vectors bear smaller packaging capacity compared to other viruses and requires several weeks to reach the maximal level of gene expression. Additionally, the scale of AAV production and the purification processes remains a challenge, though significant progress has been made over the last years [95, 121].

Considering AAV transduction, mice MSC engineered to express IFN-α were shown to be effective for the treatment of lung metastasis in an experimental model of metastatic melanoma. Treated animals showed enhanced apoptosis with a reduction in proliferation and tumour vasculature [122].

Within the context of cirrhosis, in the study by Jin and colleagues [123], rat BM-MSCs were designed to overexpress the *BCL-2* gene, which has anti-apoptotic features and can help cell survival and proliferation, to test their therapeutic effects in a rat model of liver cirrhosis. After isolating, expanding and transducing rat BM-MSCs, rats with cirrhosis induced by carbon tetrachloride were treated with *BCL-2*-expressing MSCs. Overall, the results demonstrated that the engineered MSCs showed better survival, enhanced differentiation into hepatocyte-like cells, and appeared to promote the recovery of liver function in rats with experimental cirrhosis [123].

Aiming to treat avascular necrosis of the femoral head, Liao and co-workers [124] developed lines of rat MSCs expressing either a combination of VEGF and bone morphogenetic protein-6 (BMP-6) genes, or each of these genes individually. Transduced MSCs were incorporated into a biomimetic synthetic scaffold of poly lactide-co-glycolide (PLGA) and were then subcutaneously implanted into nude mice. The results revealed that there was a significant increase in the bone volume, and a greater number of blood vessels of implants that carried BM-MSCs expressing both VEGF and BMP-6 compared to those expressing either of the growth factors alone.

Similar to Farrell's [103] and Tian's [104] studies, Nakajima and colleagues [125] also developed MSCs overexpressing IL-10, but in this study the therapeutic target was acute ischemic stroke in a rat model of middle cerebral artery occlusion (MCAO). Beyond the aforementioned effects of IL-10 in the previous section, this cytokine is also a contributing factor to neuroprotection in MSCs and suppresses expression of various pro-inflammatory cytokines. Animals were intravenously administrated with human MSCs alone or IL-10-engineered MSCs. The results demonstrated that IL-10 overexpression enhances the neuroprotective effect of MSCs and extends the therapeutic time window of MSC administration after transient focal ischemia. IL-10-overexpressing MSCs significantly reduced infarct volume, enhanced motor functional recovery and significantly reduced neuronal damage in the cortical internal border-zone (IBZ) [125].

**Table 2. Genetically modified MSCs tested for the treatment of different conditions employing viral gene delivery methods**

| Transgene | MSC source | Transduction Method | Therapeutic target | Observations | Reference |
|---|---|---|---|---|---|
| Apelin | Human Wharton' Jelly | Lentivirus | Type 2 diabetes mellitus | | [75] |
| BCL-2 | Rat Bone Marrow | AAV | Hepatic Cirrhosis | | [123] |
| bFGF | Rabbit Bone Marrow | Lentivirus | Bone regeneration | | [78] |
| bFGF & PDGF-BB | Human Placenta | Adenovirus | Hindlimb ischemia | | [105] |
| BMP-2 | Human Umbilical Cord | Lentivirus | Bone regeneration | | [76] |
| | Mouse Bone Marrow | AAV | Osteopenia | | [127] |
| BMP-6 & VEGF | Rat Bone Marrow | AAV | Bone regeneration | MSCs loaded onto a scaffold | [124] |
| BMP-7 | Canine Adipose Tissue | Lentivirus | Bone regeneration | MSCs transplanted in cell sheets | [77] |
| CNP | Rat Bone Marrow | Adenovirus | Articular cartilage regeneration | MSCs loaded onto a scaffold | [119] |
| CXCR4 | Not provided | Retrovirus | Improve MSC homing towards infarcted myocardium | | [61] |
| Decorin | Human Umbilical Cord | Adenovirus | Radiation-induced lung injuries | | [112] |
| ERK1/2 | Rat Bone Marrow | Lentivirus | Stroke | | [129] |
| HSV1-TK | Mouse Bone Marrow | Retrovirus | Anaplastic thyroid cancer | Prodrug suicide gene therapy system | [58] |
| IL-10 | Human Bone Marrow | Adenovirus | Osteoarthritis | | [103] |
| | Rat Bone Marrow | Adenovirus | Rheumatoid arthritis | | [104] |
| | Human Bone Marrow | AAV | Acute Ischemic Stroke | | [125] |
| INF-α | Mice Bone Marrow | AAV | Melanoma | | [122] |

| Transgene | MSC source | Transduction Method | Therapeutic target | Observations | Reference |
|---|---|---|---|---|---|
| Interferon-β | Human Umbilical Cord | Lentivirus | Lung cancer | | [68] |
| | Canine Adipose Tissue | Lentivirus | Melanoma | Combination of cell-based therapy and chemotherapy | [69] |
| | Bone Marrow | Lentivirus | Ovarian cancer | | [70] |
| Interleukin-12 | Bone Marrow | Retrovirus | Melanoma | | [56] |
| Interleukin-32 | Mouse Bone Marrow | Adenovirus | Lung cancer Gastric cancer | | [97] |
| Klotho | Mice Bone Marrow | Adenovirus | Acute kidney injury | | [114] |
| LIGHT | Human Umbilical Cord | Lentivirus | Gastric cancer | | [71] |
| miR-124a | Not provided | Lentivirus | Glioma | Exosome delivery | [88] |
| miR-26a | Mouse Bone Marrow | Lentivirus | Bone regeneration | | [87] |
| Netrin-1 | Mouse Adipose Tissue | Adenovirus | Diabetic peripheral neurovascular disease | | [110] |
| OPG | Human Umbilical Cord | Adenovirus | Osteosarcoma | | [98] |
| PAI-1 | Human Endometrium | Lentivirus | Decidualization insufficiency | Engineering by CRISPR-Cas9 technology | [91] |
| PEDF | Mouse Bone Marrow | Adenovirus | Lewis lung carcinoma | | [99] |
| shRNA for PHD2 | Rat Bone Marrow | Lentivirus | Periodontitis | Tissue-engineered construct to deliver MSCs | [89] |
| TNF | Human Umbilical Cord | Lentivirus | Gastric cancer | | [72] |
| TNFR | Rat Bone Marrow | AAV | Acute myocardial infarction | | [128] |

**Table 2. (Continued)**

| Transgene | MSC source | Transduction Method | Therapeutic target | Observations | Reference |
|---|---|---|---|---|---|
| TRAIL | Human Adipose Tissue | Lentivirus | Pancreatic cancer | | [82] |
| | Rat Wharton' Jelly | Lentivirus | Hepatocellular cancer | | [80, 81] |
| | Human Bone Marrow | Lentivirus | Lung cancer | | [79] |
| | Human Adipose Tissue | Retrovirus | Cervical carcinoma Pancreatic cancerColon cancer | | [52] |
| | Human Adipose Tissue | Lentivirus | Pancreatic cancer | Combination of cell-based therapy and chemotherapy | [83] |
| | Human Gingiva | Lentivirus | Tongue squamous cell carcinoma | | [84] |
| | Human Bone Marrow | Lentivirus | Colorectal carcinoma | | [85] |
| | Human Umbilical Cord | Lentivirus | Non-Hodgkin's lymphoma | | [86] |
| | Human Bone Marrow | Adenovirus | Pancreatic ductal adenocarcinoma | MSCs deliver oncolytic viruses carrying the TRAIL transgene | [101] |
| yCD::UPRT | Human: Adipose Tissue Bone Marrow Dental Pulp Menstrual Blood Umbilical Cord | Retrovirus | Prostate cancer Breast adenocarcinoma Glioblastoma | Prodrug suicide gene therapy system Exosome delivery | [57] |

As previously stated, many genes in MSCs are involved in bone formation, and bone morphogenetic protein 2 (BMP-2) is one of the well-studied factors that determines the osteoblast-biased differentiation of MSCs [126]. Kumar and co-workers [127] engineered mice MSCs, employing an AAV vector, to overexpress BMP-2 to enhance the regeneration and bone density in an immunocompetent mouse model of osteopenia. Compared to MSCs without genetic modification, bone repaired upon administration of (BMP-2)-expressing MSCs had a higher mineral density, microstructure and activity, with a relatively superior proliferative capability compared non-engineered MSCs and a higher osteocompetent pool of cells.

TNF-$\alpha$ is a cytokine that is closely related with autoimmune diseases, such as rheumatoid arthritis and myocarditis. Bao and colleagues [128] modified rat MSCs to express TNF-$\alpha$ receptor (TNFR) and used TNFR-MSCs to treat inflammation and cardiac dysfunction following acute myocardial infarction. The main objective of this study was to determine if by infusing cells expressing TNFR it could be possible to decrease gene expression and protein production of some crucial inflammatory cytokines. After infusion of TNFR-MSCs, expression of TNF-$\alpha$, IL-6 and IL-1$\beta$ was reduced, and the apoptosis of cardiomyocytes was inhibited when compared to non-modified MSCs [128].

## 3.2. Non-Viral Methods

As aforementioned, viral systems are still the conventional method to introduce a therapeutic gene into mammalian cells, owing to their higher gene transfer efficiency and long-term stable gene expression both *in vitro* and *in vivo*. However, the clinical application of virally engineered cells is still hampered due to safety issues related to risks of carcinogenesis and immunogenicity, as well as due to limited DNA packaging capacity and difficulty of vector production [130]. These limitations have motivated the development of alternative non-viral gene delivery systems, which can be classified as chemical and physical methods. The non-viral methods, featuring low immunogenicity, no risk of transmission of infectious diseases, flexibility in loaded DNA capacity and low production cost, represent a promising and effective approach currently used for gene delivery [131–133]. However, both physical and chemical methods present some drawbacks, namely lower transfection efficiencies compared to viral vectors, the possibility to disrupt cellular/nuclear membranes and the unsuitability for the transfection of a large population of cells and hard-to-transfect cells. Particularly, the physical methods are difficult to be applied in an *in vivo* situation and can lead to tissue damage. On the other hand, chemical agents can lead to toxicity at higher concentrations, cause adverse reactions with negatively charged molecules and raise safety concerns due to the undegradable nature of certain materials [133, 134]. Still, in the last decades, numerous non-viral gene delivery strategies have been developed and applied for the transfection of human cells, in particular MSCs. Examples of studies that used non-viral systems to genetically engineer human MSCs in order to adjust their therapeutic potential will be presented in the following sections.

### 3.2.1. Chemical Methods

Chemical methods imply the use of natural or synthetic materials, also referred to as nanocarriers, which electrostatically condense or encapsulate nucleic acids into nanoparticles (NPs) or aggregate complexes that subsequently transfer the genetic material into the cell. These delivery systems can be physicochemically tuned in order to overcome the barriers and engineered to target specific tissues for improved gene delivery and expression [135]. A wide variety of nanocarriers including polymers, lipids, polysaccharides, peptides, and inorganic materials, have been used to facilitate transfection of MSCs [133, 134]. A summary of several studies where MSCs were genetically engineered using nanocarriers is shown in Table 3.

Cationic lipid- and polymeric- based nanocarriers, such as lipofectamine and polyethylenimine (PEI), are considered the gold standard for non-viral gene transfection and often used as transfection control, including studies specifically employing MSCs. For example, Cho and co-workers [136] tested the delivery of a vector encoding the human forkhead box A2 (*Foxa2*) gene to rat BM-MSCs using Lipofectamine, aiming at stimulating of tissue regeneration after cell administration, protecting the liver from hepatic diseases. In this study, *Foxa2*-engineered MSCs showed to efficiently incorporate into liver grafts.

Also using the lipofection method, Tsubokawa and colleagues [137] transiently overexpressed the human heme oxygenase-1 (HO-1) gene, which encodes an anti-oxidant and anti-inflammatory protein with potential to attenuate ischemic myocardial injury, in rat BM-MSCs. The results showed that the engineered MSCs exhibited enhanced anti-apoptotic and anti-oxidative abilities, contributing to improved repair of ischemic myocardial injury in a rat infarction model, through cell survival and VEGF production. Additionally, lipofectamine has also been used to engineer human BM-MSCs to express bacterial cytosine deaminase (CD), which catalyses the hydrolytic deamination of the non-toxic 5-FC molecule into the anticancer drug 5-FU. CD-expressing MSCs showed anticancer therapeutic potential while minimizing the side-effects of 5-FU. *In vivo* studies have demonstrated that by intravenous injection, followed by systemic administration of 5-FC, the engineered cells are able to inhibit the growth of human gastric tumours in mice [138].

Furthermore, RNAi molecules have also been successfully transfected into MSCs using lipofectamine, for example in the study by Yu and collaborators [139], mouse BM-MSCs were transfected with let-7a miRNA inhibitors. let-7a is a miRNA that targets the mRNA of Fas and Fas Ligand (FasL), which are crucial proteins in the context of MSC therapy for inflammatory diseases, as these enhance the migration of T cells and activate the apoptosis pathway in T cells, respectively. The results showed that knockdown of let-7a significantly promoted MSC-induced T cell migration and apoptosis *in vitro* and *in vivo* while reducing mortality, suppressing the inflammation reaction, and alleviating the tissue lesion of experimental colitis and GvHD mouse models.

Moreover, synthetic siRNA molecules were also successfully delivered to human BM-MSCs using lipofectamine, as described by Teoh and colleagues [140]. BM-MSCs promote the growth of myeloma cells mainly through IL-6 secretion, thus targeting its overexpression should disrupt the favourable microenvironment provided by the BM for multiple myeloma cell growth. siRNA-IL-6-transfected MSCs inhibited cell growth and IL-6 production by the human multiple myeloma cell line U266 *in vitro*. Importantly, the modified MSCs led to tumours with significantly reduced volumes and mitotic indexes in a murine subcutaneous model of human multiple myeloma.

On the other hand, Li and colleagues [141] used PEI to deliver a plasmid encoding the antiapoptotic gene *BCL-2* to rat BM-MSCs, as an attempt to improve functional recovery after acute myocardial infarction with BCL-2-engineered MSCs. *In vivo* studies showed an increased cellular survival and capillary density in the infarct border zone as well as a remarkable functional recovery upon administration of *BCL-2*-engineered MSCs into rats models of myocardial infarction.

In the study by Rejman and co-workers [142], the authors tested the cationic nanocarriers Lipofectamine, DOTAP/DOPE, and PEI for their ability to mediate the delivery of mRNA encoding the CXCR4 receptor into mouse MSCs, with the aim of increasing their homing to injured tissues. mRNA complexed with the cationic lipids transfected a larger number of cells than the polyplexes, obtaining 80% and 40% of CXCR4-positive cells, respectively. Overall, it was possible to demonstrate that mRNA could be a suitable alternative to plasmid DNA in MSCs. Future studies should focus on confirming functionality of the engineered regarding their homing capacity.

More complex polymers have also been developed and used to transfected MSCs. For example, Park and collaborators [143] used PEI-modified PLGA nanoparticles to transfect human BM-MSCs with plasmids encoding the exogenous SRY-box (SOX) *trio* (SOX 5, 6, and 9) genes, which successfully led to increased chondrogenesis of MSCs in *in vitro* culture systems.

In the context of therapeutic angiogenesis, Yang and colleagues [144] developed a set of biodegradable poly (β-amino esters) (PBAE) nanocarriers capable of delivering a VEGF-expressing plasmid to human BM-MSCs. The engineered cells led to 2- to 4-fold higher vessel densities in the subcutaneous model and also enhanced angiogenesis and limb salvage after intramuscular injection into mouse ischemic hindlimbs. As such, the engineered MSCs might represent a promising therapeutic tool for vascularizing tissue constructs and treating ischemic disease.

Considering the aforementioned potential of MSCs as tumour-targeting gene delivery vehicles, in Malik et al. (2018) [145], the authors used a PEI-polylysine (PLL) copolymer for the delivery of the exogenous suicide gene HSV-TK and also TRAIL genes to rat BM-MSCs, as a potential combinational suicidal gene therapy for glioblastoma. After intratumoural injection, the double-transfected MSCs along with prodrug GCV administration, induced a significant synergistic therapeutic response both *in vitro* and *in vivo* in a rat C6 glioma model. Likewise, Zhang and colleagues [146] used a spermine-pullulan (SP) copolymer to engineer rat BM-MSCs to transiently express HSV-TK, in order to investigate their effect on pulmonary melanoma metastasis. The engineered cells combined with the prodrug GCV, showed significantly tumour growth inhibition both *in vitro* and *in vivo* in the metastasis tumour model used.

In a different context, Huang and colleagues [147] also used SP for the transfection of peptide-modified rat BM-MSCs with miR-133b that is known to promote functional recovery from cerebral ischemia. The miR-133b-transfected MSCs were shown to be beneficial for the recovery of injured neural cells as the administration of engineered cells resulted in a significant increase of the number of surviving astrocytes after subjection to oxygen-glucose deprivation (OGD) *in vitro*, suggesting a therapeutic role of the modified cells to increase cell survival, thus enhancing the recovery of ischemic injured cells.

Inorganic materials have also been used for gene delivery into MSCs including gold nanoparticles (AuNPs), magnetic nanoparticles (mNPs) and silica. Aiming at the development of an effective gene delivery system for bone regeneration, Kim and colleagues [148] used mesoporous silica modified with amine groups as nanocarriers to deliver the *BMP2* gene, which encodes a bone morphogenetic protein important in bone repair and regeneration, into primary rat BM-MSCs. Also using silica, Zhu and colleagues [149] used hollow mesoporous organosilica nanoparticles conjugated with PEI to facilitate the delivery of the *HGF* gene into rat BM-MSCs. The engineered cells were transplanted into rat models of myocardial infarction and shown to be efficient in cardiac repair by largely decreasing apoptotic cardiomyocytes, reducing infarct scar size, and increasing angiogenesis in myocardium.

Aiming to improve the transfection efficiency of difficult-to-transfect MSCs, Das and co-workers [150] used AuNPs, which currently have diverse biomedical applications, modified with PEI. PEI-entrapped gold nanoparticles and covalently bound PEI-gold nanoparticles were tested as potential vehicles for the delivery of the CCAAT/enhancer binding protein beta (C/EBP-β) gene, which encodes a key transcriptional regulator of adipogenic differentiation into human WJ-MSCs. Overexpression of exogenous C/EBP-β significantly enhanced adipogenesis in MSCs, while the nanoparticles/DNA complexes showed favourable cytocompatibility in cell viability assays. This study showed that theses NPs represent a promising vehicle for gene delivery to control MSC differentiation and potentially other therapeutic gene delivery applications. Also using AuNPs, Muroski and colleagues [151] developed a novel zwitterionic cell penetrating pentapeptide that upon conjugation with AuNP facilitated the delivery of a linearised plasmid encoding for a brain-derived neurotrophic factor (BDNF) to rat BM-MSCs.

In the context of neural repair, Wu and collaborators [152] produced mNPs using synthetic hydroxyapatite (Hap) and natural bone mineral (NBM) capable of delivering a plasmid encoding glial cell line-derived neurotrophic factor (GDNF), a potent neurotrophic growth factor, into rat BM-MSCs under the action of a magnetic field. The results showed that these mNPs could be used for safe and effective transfection *in vitro*, and MSC *ex vivo* transfected with the exogenous *GDNF* might be a promising cell therapy for the treatment of neurodegenerative diseases.

**Table 3. Genetically modified MSCs targeting different therapeutic settings, using different chemical non-viral gene delivery methods**

| Nanocarrier | MSC source | Therapeutic Cargo | Application | Reference |
|---|---|---|---|---|
| Lipofectamine | Rat Bone Marrow | Forkhead box A2 (Foxa2) encoding plasmid | Regeneration of damaged liver tissues | [136] |
| | Rat Bone Marrow | Heme oxygenase-1 (HO-1) encoding plasmid | Treatment of myocardial ischemia | [137] |
| | Human Bone Marrow | Cytosine deaminase (CD) encoding plasmid | Target human gastric tumours | [138] |
| | Mouse Bone Marrow | miRNA targeting miR-let-7a | Reduced immunogenicity of MSC transplants | [139] |
| | Human Bone Marrow | siRNA targeting interleukin-6 (IL-6) | Treatment of multiple myeloma | [140] |
| Polyethylenimine (PEI) | Rat Bone Marrow | B-cell lymphoma 2 (Bcl-2) encoding plasmid | Functional recovery after acute myocardial infarction | [141] |
| Lipofectamine; DOTAP/DOPE; PEI | Not provided | C-X-C chemokine receptor type 4 (CXCR4) mRNA | Increase homing | [142] |
| PEI-modified poly lactic-co-glycolic acid (PLGA) | Human Bone Marrow | SOX-5, -6, and -9 encoding plasmids | Increase chondrogenesis | [143] |
| Poly (β-amino esters) (PBAE) | Human Bone Marrow | Vascular endothelial growth factor (VEGF) encoding plasmid | Treatment of ischemic diseases | [144] |
| PEI-modified polylysine (PLL) | Rat Bone Marrow | HSV-TK and tumour necrosis factor-related apoptosis-inducing ligand (TRAIL) encoding plasmid | Target glioblastoma | [145] |
| Spermine-Pullulan (SP) | Rat Bone Marrow | Herpes simplex virus thymidine kinase (HSV-TK) encoding plasmid | Target pulmonary metastasis | [146] |
| SP | Rat Bone Marrow | miRNA-133b | Treatment of cerebral ischemia | [147] |

## Table 3. (Continued)

| Nanocarrier | MSC source | Therapeutic Cargo | Application | Reference |
|---|---|---|---|---|
| Nanocarrier | MSC source | Therapeutic Cargo | Application | Reference |
| Silica modified with amine groups | Rat Bone Marrow | Bone morphogenic protein 2 (BMP2) encoding plasmid | Bone repair and regeneration | [148] |
| Hollow mesoporous organosilica conjugated with PEI | Rat Bone Marrow | Hepatocyte growth factor (HGF) encoding plasmid | Cardiac repair | [149] |
| PEI-coated gold nanoparticles (AuNP) | Human Wharton' jelly | CCAAT/enhancer binding protein (C/EBP) beta encoding plasmid | Increase adipogenesis | [150] |
| Zwitterionic pentapeptide conjugated with the AuNP | Rat Bone Marrow | Brain-derived neurotrophic factor (BDNF) encoding linearized plasmid | Treatment of Neurodegenerative diseases | [151] |
| Magnetic hydroxyapatite (Hap) nanoparticles; Natural bone mineral (NBM) | Rat Bone Marrow | Glial cell line-derived neurotrophic factor (GDNF) encoding plasmid | Treatment of Neurodegenerative diseases | [152] |
| PEI; Hap; RALA peptide. | Porcine Bone Marrow | Transforming growth factor beta 3 (TGF-β3) and BMP2 encoding plasmids | Increase chondrogenesis; Increase osteogenesis | [153] |

It is important to note that MSCs response following non-viral gene transfection may strongly differ depending on the selected nanocarrier, as reported in the study by Gonzalez-Fernandez co-workers [153]. The authors tested three non-viral gene delivery systems: PEI, Hap and an amphipathic peptide RALA to transiently deliver *TGFβ3* and/or *BMP2* genes to porcine BM-MSCs in order to promote their osteogenesis or chondrogenesis and evaluate the influence of different gene nanocarriers on the lineage commitment of MSCs. Despite showing similar gene transfection efficiencies these nanocarriers had different effects on MSC viability, cytoskeletal morphology and differentiation, as transfection of MSCs using PEI failed to induce robust osteogenesis or chondrogenesis, whereas Hap transfection promoted osteogenesis in 2D culture and RALA transfection showed to be less osteogenic and promote a more cartilage-like phenotype, confirmed through relative expression levels of chondrogenic markers Aggrecan (ACAN) and SOX- 9 and osteogenesis marker Runt related transcription factor 2 (Runx2), and biochemical and histological analysis of GAG, calcium and collagen. The results demonstrate that the differentiation of MSCs through the application of non-viral gene delivery strategies also depends on the nanocarrier itself and not only on the gene delivered.

### 3.2.2. Physical Methods

Physical methods often imply *ex vivo* transfection of cells by disruption of cellular/nuclear membranes and consequent transfer of nucleic acids. Among different methods, microinjection, particle bombardment, electroporation, sonoporation and laser irradiation have been used to facilitate gene transfer into mammalian cells [154]. Some of these methods have been used to genetically engineer MSCs for enhanced therapy, including electroporation and sonoporation. A summary of different studies of genetic modification of MSCs using physical methods is shown in Table 4.

**Table 4. Genetically modified MSCs targeting different therapeutic settings, using different physical non-viral gene delivery methods**

| Method | MSC source | Therapeutic Cargo | Application | Reference |
|---|---|---|---|---|
| Electroporation | Human Bone Marrow | SOX 5, 6, and 9 encoding plasmids | Increase chondrogenesis | [155] |
| | Human Adipose Tissue | Runx2 and Osterix encoding plasmids | Increase osteogenesis | [156] |
| | Human Bone Marrow | STEAP3, syndecan-4 (SDC4) and L-aspartate oxidase (NadB) encoding plasmid | Increase exosome production | [157] |
| | Human Umbilical Cord | Heat shock protein 27 (Hsp27) encoding plasmid | Neurological recovery after stroke | [158] |
| Method | MSC source | Therapeutic Cargo | Application | Reference |
| | Human: Adipose Tissue Umbilical Cord Blood Bone Marrow | Beta 2-microglobulin (B2M) gene knock-out with *CRISPR-Cas9* system | Reduced immunogenicity of allogeneic transplants | [159] |
| Nucleofection | Human Umbilical Cord Blood | PDX-1 mRNA | Treatment of diabetes mellitus | [161] |
| | Human Adipose Tissue | Tumour necrosis factor-related apoptosis-inducing ligand (TRAIL) encoding plasmid | Target cancer | [162] |
| | Human Adipose Tissue | Neurotrophins-3 (NT-3) encoding plasmid | Target medulloblastoma | [163] |
| | Porcine Adipose Tissue | Bone morphogenetic protein-6 (BMP-6) encoding plasmid | Vertebral bone repair | [164] |
| | Umbilical Cord Blood | Soluble Receptor for AGEs (sRAGE) expression with *CRISPR-Cas9* system | Treatment of Parkinson's Disease | [165] |

| Method | MSC source | Therapeutic Cargo | Application | Reference |
|---|---|---|---|---|
| Microporation | Human Bone Marrow | C-X-C chemokine receptor type 4 (CXCR4) encoding minicircle | Increase homing | [167] |
| Method | MSC source | Therapeutic Cargo | Application | Reference |
| Microporation | Human Bone Marrow | Vascular endothelial growth factor (VEGF) encoding minicircle | Treatment of peripheral artery disease | [168] |
| Sonoporation | Bovine Dental Pulp | Growth/differentiation factor 11 (Gdf11) encoding plasmid | Dental tissue repair | [169] |
| | Rat Bone Marrow | siRNA targeting phosphatase and tensin homolog on chromosome 10 (PTEN) | Increased viability after transplantation | [170] |
| | Rat Bone Marrow | Hemopexin-like domain fragment (PEX) encoding plasmid | Target cancer | [175] |
| | Rat Bone Marrow | CXCR4 encoding plasmid | Engraftment to acute kidney injury (AKI) tissues | [172] |

Electroporation is one of the most common non-viral gene delivery method and relies on the exposure of the cell membrane to high-intensity electrical pulses. The transient and localized destabilization of the barrier allow the permeabilization of exogenous molecules, such as DNA, present in the surrounding medium [154]. Electroporation is cost-effective and an efficient method that is widely used for *ex vivo* transfection of MSCs [134].

For example, Kim and colleagues [155] used electroporation to facilitate the delivery of a plasmid encoding the SOX trio genes into human BM-MSCs. The results showed that electroporation-mediated co-transfection of these transcription factors enhances chondrogenesis and suppresses hypertrophy of MSCs. On the other hand, Lee and co-workers [156] reported that the electroporation-mediated transfection of plasmid encoding the Runx2 and Osterix transcription factors, which are essential for osteogenic differentiation from uncommitted progenitor cells, into human AT-MSCs enhances their in vitro and *in vivo* osteogenesis.

Kojima and collaborators [157] used electroporation to engineer human BM-MSCs to increase exosome production, which are promising therapeutic agents, by facilitating the delivery of a plasmid encoding for STEAP3, syndecan-4 (SDC4) and L-aspartate oxidase (NadB) proteins that are involved in exosome biogenesis. The combined expression of these genes from the engineered MSCs led to significantly increased exosome production, up to a 40-fold increase compared to non-modified MSCs. This booster effect was detected by supernatant luminescence due to the expression of a construct encoding the vesicle-associated CD63 marker fused with nanoluc that was co-transfected into MSCs.

Another study that used electroporation as delivery system was reported by Liu and colleagues [158], in which human UC-MSCs were transfected with a vector encoding the non-senescent heat shock protein 27 (Hsp27). Cells undergoing replicative senescence have diminished heat shock response, thereby alterations in Hsp7 production can potentially increase the accumulation of damaged proteins promoting further aging. The results showed that the engineered MSCs promoted neuroplasticity in a mouse stroke model and decreased cellular senescence, thus increasing MSC engraftment.

Electroporation has also been used for the delivery of CRISPR-Cas9-mediated genome editing systems, for example in the study Xu and colleagues [159], where tube electroporation was used to deliver a CRISPR-Cas9 system that generates gene knock-out via precise homology-directed repair (HDR) to MSCs from different human sources. The authors co-delivered a Cas9/gRNA ribonucleoprotein with a sequence targeting the beta 2-microglobulin (B2M) gene, which is a well-established MHC class I molecule association protein, and a small DNA repair template that introduces a single base insertion in the PAM sites creating a frameshift mutation that disrupts gene function. The delivery of such system resulted in B2M gene disruption, showing a maximum reduction in the cell-surface B2M expressing cells of 80% [159].

Alternatively to conventional electroporation, the commercial system Nucleofector (Lonza, Germany) has also been used to efficiently transfect MSCs, by employing cell-type specific pulse parameters and buffer formulations that allows the direct electroporation of DNA into the nucleus – nucleofection [160]. Pham and colleagues [161] used nucleofection to transfect human UCB-MSCs with mRNA encoding PDX-1, a transcription factor also known as insulin promoter factor 1, whose expression has been known to facilitate MSCs differentiation into insulin-producing cells (IPCs). The results showed that PDX-1 mRNA transfection significantly improved the differentiation of MSCs into IPCs and led to 2-fold increase in production of insulin and C-peptide in response to glucose. Therefore, this approach could be a promising system to produce safe IPCs as a potential therapy for diabetes mellitus treatment.

The nucleofection method was also used by Fakiruddin and collaborators [162] to facilitate the delivery of TRAIL-encoding vectors into human AT-MSCs, in order to evaluate the anti-tumourigenic potential of the engineered MSCs *in vitro* using several cancer models. TRAIL-expressing MSCs selectively inhibit both tumour and malignant lines proliferation and markedly induces apoptosis [162]. In another example of a potential anti-cancer therapy, Kim and colleagues [163] engineered human AT-MSCs to target medulloblastoma (MBL), the common malignant childhood brain tumour, using nucleofection as transfection method. AT-MSCs were transfected with plasmid encoding neurotrophins-3 (NT-3), found to induce tumour cell apoptosis in MBL. The transfected cells showed a growth-inhibitory effect on a MBL cell line *in vitro*, by inducing apoptotic tumour cell death and neuronal differentiation of tumour cells, showing potential as a targeted gene therapy for MBL.

Also using nucleofection, but in the context of bone repair, Pelled and colleagues [164] genetically modified porcine AT-MSCs to transiently overexpress BMP-6 protein as a potential therapy for a variety of conditions involving bone loss, namely osteoporotic vertebral compression fractures. In this study, the engineered cells showed to induce vertebral defect regeneration in a clinically relevant, large animal pig model.

More recently, nucleofection was used to transfect a CRISPR-Cas9 ribonucleoprotein system in order to insert an expression cassette into UCB-MSCs [165]. The insert consisted of human elongation factor 1-alpha (EF1-α) promoter, soluble Receptor for AGEs (sRAGE) coding sequence and poly A tail. sRAGE, is an inhibition factor of Advanced Glycation End products- albumin (AGE-albumin) from activated microglial cells, which is one of the main causes of Parkinson's disease (PD). After transplantation into a PD mouse model, MSCs-expressing sRAGE showed to extensively reduce neuronal cell death in Corpus Striatum and Substantia Nigra and improved movement recovery [165].

Although effective at transfection, cuvette-based electroporation is limited by cytotoxicity attributed effects of the pulsed electric fields on biomolecules, pH variation, increasing in temperature, and metal ion generation. Microporation is a unique electroporation technology that uses a pipette tip as an electroporation space and a capillary type of electric chamber instead of a cuvette, counteracting the harmful effects of conventional cuvette-based methods [166]. This technique has also been used in studies specifically employing MSCs, for instances, to facilitate the delivery of a minicircle vector encoding the *CXCR4* gene into human BM-MSCs using microporation [167]. In this particular study, the engineered cells showed greatly increased their *in vivo* homing ability towards injury sites in a mouse model, compared to non-modified cells [167]. Recently, Serra and co-workers [168] also used microporation to deliver a minicircle vector encoding the VEGF gene into human BM-MSCs. VEGF-overexpressing MSCs showed an improved angiogenic potential *in vitro*, confirmed by endothelial cell tube formation and cell migration assays.

Sonoporation relies on the application of ultrasounds to transiently enhance cell permeability through formation of small pores in the membrane, allowing for the direct transfer of genetic material into cells. The mechanism involved in sonoporation appears to be acoustic cavitation that through mechanical perturbation, collapse of active bubbles and the associated energy release leads to the permeabilization of adjacent cell membranes [154]. Although very few MSC-related studies have been reported, Nakashima and colleagues [169] used sonoporation for the transfection of a plasmid encoding the growth/differentiation factor 11 (Gdf11), a member of the BMP family expressed in terminally differentiated odontoblasts, into bovine dental pulp-derived MSCs. After *in vitro* transfection, increased expression of dentin sialoprotein, a differentiation marker for odontoblasts, was shown, suggesting differentiation of the pulp stromal MSCs into odontoblasts. Additionally, after *in vivo* transfection, restoration of the amputated dental pulp was observed in canine teeth.

In another study, sonoporation using a combination of ultrasound and microbubbles was used to deliver siRNAs into rat BM-MSCs in order to silence the expression of phosphatase and tensin homolog on chromosome 10 (PTEN), which is known to be a tumour suppressor gene that negatively regulates PKB/Akt-dependent cell survival. The transfected cells showed reduced levels of PTEN-mRNA and increased levels of Akt phosphorylated protein. Thus, PTEN inactivation could be considered a promising method to improve the viability and therapeutic efficacy of transplanted MSCs [170]. Also in the context of anti-cancer therapies, more recently, Haber and co-workers [171] used ultrasound-mediated transfection to engineer MSCs to target cancer. To this end, rat BM-MSCs were transfected with a plasmid encoding for hemopexin-like domain fragment (PEX), an inhibitor of tumour angiogenesis. The results showed that PEX-secreting MSCs is a promising cell-based delivery approach in cancer settings, by inhibiting prostate tumour growth up to 70% following a single I.V. administration and up to 84% after repeated administration to mice bearing prostate tumours.

In the context of kidney repair, and envisaging the therapeutic benefits of an enhanced homing capacity by MSCs, Wang and colleagues [172] used microbubble-mediated ultrasound combined with PEI to facilitate the delivery of a vector encoding *CXCR4* into rat BM-MSCs in order to improve their homing towards AKI. After injected into rats' tail veins, the modified MSCs showed enhanced homing and retention into AKI-induced tissues.

Finally, microinjection is also considered a promising gene delivery method that uses a glass needle to directly introduce the genetic material into the cell cytoplasm or nucleus, by hydrostatic pressure. The injection is carried out on a single cell under direct visual control, using a microscope. Although conceptually simple, microinjection is difficult to apply since it requires extreme precision and it is impractical for transfecting large numbers of cells. Although this method is not commonly employed for MSC engineering, microinjection with nanoneedles has proven to be efficient in delivering plasmid DNA into MSCs, showing 65–75% reporter transgene expression efficiency while retaining cell viability after injection [173, 174].

## 4. CLINICAL TRANSLATION OF GENETICALLY ENGINEERED MSCS

As previously detailed, MSCs possess a variety of therapeutic features including their differentiation potential, homing capacity, immunomodulatory activity, trophic/paracrine effects and transfer of vesicular components. Therefore, MSCs have been considered as a potential cell population to develop therapies for various human disorders including cancer, cardiovascular and ischemic diseases, as well as starting cell material for tissue engineering strategies. Although MSC-based therapy has showed to be safe and effective in various clinical trials, there are still several limitations that restrict their clinical implementation. Hence, *ex vivo* gene delivery to MSC (i.e., *ex vivo* MSC-based gene therapy) has been used to enhance the therapeutic efficacy and clinical potential of these cells. In this section, a selection of clinical trials testing cell therapies with genetically engineered MSCs are described and summarized in Figure 3 and Table 5 (information obtained at "clinicaltrials.gov" on May 15th 2019, using the terms "mesenchymal stem cells gene"; "mesenchymal stem cells viral"; and "mesenchymal stem cells cancer").

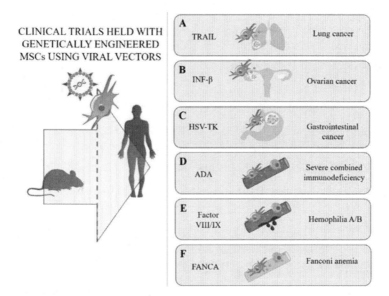

Figure 3. Clinical trials conducted with engineered mesenchymal stromal cells (MSCs) using viral vectors as a gene delivery method, targeting different pathologies. A) tumour necrosis factor-related apoptosis-inducing ligand (TRAIL) -expressing MSCs, targeting lung cancer (NCT03298763); B) Interferon-β (INF-β) –expressing MCSs, targeting ovarian cancer (NCT02530047); C) herpes simplex virus-1 thymidine kinase (HSV-TK) –expressing MSCs, targeting gastrointestinal cancer (NCT02008539); D) adenosine deaminase (ADA) expressing MSCs and HSCs, to treat severe combined immunodeficiency (SCID); E) human clotting factor VIII or factor IX – expressing MSCs and HSCs, to treat haemophilia A (NCT03217032) or B (NCT03961243), respectively; and F) FANCA-expressing MSCs and HSCs, to treat Fanconi anemia (NCT02530047).

Taking advantage of their tumour tropism characteristics, MSCs have been widely used in preclinical studies as a vehicle for the delivery of anti-cancer agents, supporting their investigation at the clinical level. For example, the completed phase I/II study, held in Germany until 2017 (NCT02008539), used MSCs genetically modified to express HSV-TK (MSC_apceth_101), using a retroviral SIN-vector, for the treatment of advanced gastrointestinal adenocarcinoma. The study was conducted with 10 patients that received three cycles of infusion of MSC_apceth_101, followed by GCV administration. Despite some adverse events were reported, five patients achieved a stable disease state and all patients showed a median time to treatment progression of 1.8 months and an overall survival of 15.6 months. Overall, MSC_apceth_101 in combination with GCV was safe and tolerated well by patients with advanced gastrointestinal adenocarcinoma, and showed preliminary signs of efficacy in terms of clinical stabilization of disease [176].

As aforementioned, MSC have been engineered to express IFN-β using lentivirus as a gene delivery system in a variety of cancer studies, including ovarian cancer [70]. In this regard, a phase 1 study, conducted in the United States since 2015 (NCT02530047 (active, not recruiting)), has been determining the effects of the administration of MSCs genetically modified by a plasmid vector to produce IFN-β (MSC-INFβ), in patients with advanced ovarian cancer. In this trial, allogenic MSC isolated from the BM of healthy male donors are being used. The main objective of this study is to determine the highest tolerable dose of MSC-INFβ that can be administered to patients with ovarian cancer and test the safety of the MSC-INFβ were infused intraperitoneally through a catheter placed in the abdomen of 21 female patients, with histologically documented diagnosis of epithelial ovarian cancer. After a specific time period, the status of the disease will be checked in patients will have through tumour biopsies.

Similarly, TRAIL-expressing MSCs appear as a promising anti-cancer strategy with pre-clinical relevance against several types of cancer models. In particular, lentivirus have been used as a gene delivery method to engineered MSC to express TRAIL, targeting lung metastatic cancer [79]. In this context, a phase I/II study conducted in the United Kingdom since

2017 (NCT03298763 (recruiting)), known as the TACTICAL, has been using MSCs engineered to express the TRAIL gene (MSCTRAIL) using a lentiviral vector, for the treatment of lung cancer. MSCTRAIL will be intravenously administered to 46 patients with lung adenocarcinoma, in combination with pemetrexed/cisplatin chemotherapy. The aim of the first part of study is to assess the recommended dose of MSCTRAIL for the subsequent phase II. The second part of the research will compare the effect of MSCTRAIL when given with chemotherapy compared to chemotherapy alone, aiming at assessing tolerability and preliminary efficacy of MSCTRAIL.

In a different context, another phase I/II study, which has been started in 2017 (NCT03351868 (recruiting)), will test the co-administration of genetically modified hematopoietic stem/progenitors cells (HSPCs) and MSCs for the treatment of Fanconi Anemia (FA), a rare, inherited disease that is caused by the defection of the FANCA gene. Hematopoietic cell transplantation (HCT) is a common therapy for this condition, however, risks associated with HCT include rejection of the transplanted cells and GVHD. In this clinical study, autologous HSPCs and MSCs will be transduced with the FANCA gene *ex vivo* with a self-inactivating lentiviral vector, for the treatment of Fanconi anemia disease by correcting the defective gene in HSPCs. The primary aim of this study is to evaluate the safety and efficacy of the *ex vivo* gene transfer clinical protocol. Gene-modified autologous cells will be infused into 30 FA patients. Also in the context of HCT, two phase I/II trials will be using genetically modified HSPCs and MSCs to express human clotting factor VIII (YUVA-GT-F801) (NCT03217032 (not yet recruiting, 10 patients estimated)) or factor IX (YUVA-GT-F901) (NCT03961243 (not yet recruiting, 10 patients estimated)), using an advanced lentiviral vector, for the treatment of hemophilia A or hemophilia B, respectively. The primary aim of both studies is to assess the safety and preliminary efficacy of the genetically modified autologous cells.

**Table 5. Representative clinical trials testing the safety and preliminary efficacy of genetically modified MSCs in different therapeutic settings**

| Study Title | Phase | State | Country | Reference | Last Update Date |
|---|---|---|---|---|---|
| Treatment of Advanced Gastrointestinal Cancer in a Phase I/II Trial With Modified Autologous MSC_apceth_101 | Phase I/II | Completed | Germany | NCT02008539 | March 27, 2017 |
| Phase 1 Study to Determine the Effects of Mesenchymal Stem Cells Secreting Interferon Beta in Patients With Advanced Ovarian Cancer | Phase I | Active, not recruiting | United States | NCT02530047 | May 9, 2019 |
| Targeted Stem Cells Expressing TRAIL as a Therapy for Lung Cancer | Phase I/II | Recruiting | United Kingdom | NCT03298763 | March 8, 2019 |
| Gene Transfer for Fanconi Anemia Using a Self-inactivating Lentiviral Vector | Phase I/II | Recruiting | China | NCT02530047 | May 17, 2018 |
| Gene Modified Hematopoietic and Mesenchymal Stem Cells for Hemophilia A and B | Phase I/II | Not yet recruiting | China | NCT03217032 NCT03961243 | May 17, 2018 |
| Gene Transfer for Adenosine Deaminase-severe Combined Immunodeficiency (ADA-SCID) Using an Improved Self-inactivating Lentiviral Vector (TYF-ADA) | Not provided | Not yet recruiting | China | NCT03645460 | October 30, 2018 |

Finally, a clinical study testing the administration of autologous HSPCs and/or MSCs transduced with a self-inactivating lentiviral vector carrying the adenosine deaminase (ADA) gene (TYF-ADA), will be held in China for the treatment of patients with severe combined immunodeficiency (SCID) due to a defective ADA gene. The study will be conducted in 10 patients diagnosed with ADA-SCID and severe infections. The primary objectives are to assess the safety TYF-ADA and preliminary efficacy of the *ex vivo* gene transfer clinical protocol towards immune reconstitution in patients overcoming frequent infections present at the time of treatment (NCT03645460 (not yet recruiting)).

## CONCLUSION

MSCs have unique therapeutic properties and hold great promise for regenerative medicine and cell therapy, with encouraging results in several pre-clinical studies or early stage clinical trials. Over the last years, significant efforts have been made to address the limitations of MSCs in early clinical trials, namely by using genetic engineering tools to improve the therapeutic potential of these cells. Despite the advantages of non-viral gene delivery methods, to date, all conducted clinical trials based on genetically engineered MSCs are relying on the use of viral methods. However, relevant progress has been made in recent years towards non-viral strategies, which opened new possibilities in the field of cell and gene therapy. Of notice, novel and precise genome-editing technologies such as zinc finger nucleases (ZFNs) [177], transcription activator-like effector nucleases (TALENs) [178] and CRISPR [179] have been recently developed that can be employed for safer site-specific gene modifications and thus induce a stable long-term transgene expression in MSCs.

There is still a number of challenges when translating the genetically modified MSCs therapy into the clinics, particularly related to the manufacturing of a clinical-grade product. The quality (i.e., potency) and quantity (i.e., cell dose) of the cellular product are the two central issues that could enable researchers to effectively compare results and develop more efficient MSC-based therapies. Cell and gene therapies are inherently based on heterogenous living products and their characteristics can be affected by multiple variables associated with their *ex vivo* manipulation, namely *ex vivo* expansion to achieve clinically meaningful cell numbers. Differences in the techniques available for the large-scale expansion of MSCs, such as the culture medium used, matrices to support cell anchorage and devices for cell propagation (i.e., bioreactors), may influence the heterogeneity, senescence, secretome and multipotent potential of the cell population of interest, as well as the potential for *in vivo* homing, survival and integration into damaged tissues. Most clinical protocols to date implement extensive MSC expansion protocols, using a variety of conditions, which may impact cell identity and phenotype, and consequently the therapeutic potential of these cells [180, 181]. Thus, a simple, streamlined manufacturing process needs to be developed, while remaining relatively cost-effective, in order to allow translation to clinical use. Moreover, although BM remains the gold standard as a MSC tissue source, cells isolated from AT and UC have also gained relevance in cell/gene therapy studies. As such, pre-clinical and clinical studies involving different methods of MSC isolation (including distinct tissue sources) and *ex vivo* expansion might lead to difficulties in comparing the studies' outcomes.

Another point of crucial importance concerns the challenges of using autologous cell-based therapies. The use of autologous MSCs has proven to be feasible and a safe approach in numerous cell-based therapy clinical studies [182]. However, it is difficult to isolate enough numbers of autologous MSCs, with high activity and low variability, from patients with advanced age or comorbidities. Furthermore, autologous MSCs are not suited for the treatment of acute diseases or conditions that require the generation of modified MSC-based therapies because their extraction and subsequent *ex vivo* expansion to produce the high number of cells required is extremely time-consuming [182]. Allogeneic MSCs, featuring immunosuppressive properties and low immunogenicity, have been studied as an alternative to autologous cells, as these can be obtained in advance from young healthy donors, expanded and kept cryopreserved in cell banks enabling their immediate off-the-shelf use. Allogeneic MSC-based therapies are thus a promising alternative to autologous cells, with advantages in what concerns time, cost, and quality assurance [182]. In this context, the ongoing clinical trial TACTICAL (NCT03298763) is aimed at manufacturing a clinical-grade MSC-based product targeting lung adenocarcinoma, while assuring their cost-effectiveness. To produce the number of cells required, allogeneic UC-MSCs from multiple donors are transduced and subsequently pooled, in order to ensure maximum production capacity and to reduce the impact of the inherent variability of MSCs [180, 183] Cell expansion is then carried out in the single use Xpansion® Multiplate Bioreactor System (Pall Corporation, US) assuring compliance with Good Manufacturing Practices (GMP). Although some literature has suggested that cryopreservation can affect certain key therapeutic characteristics of cell products [181], the cryopreservation procedure of TACTICAL has been adapted to ensure that the product is not affected [184].

Another important point regarding the translation of MSC-based therapies concerns the protocol for cell administration. Frequency, timing, route and dose of administration may impact the therapeutic benefit of MSC-based products, by affecting their *in vivo* distribution, long-term viability as well as their biological fate [181, 185]. In fact, the dosing regime is one of the crucial issues that is limiting the progress of MSC-based therapies towards routine clinical implementation. Thus, researchers and clinicians have been using pharmacokinetic modelling to characterize and predict the *in vivo* kinetics of systemically administered MSCs in order to maximize their therapeutic activity and minimize potential side effects [185]. Recent pharmacokinetic models suggest that MSCs may need to be administered at larger doses and/or more frequently in order to sustain a long-term biological response. Alternatively, through bypassing the initial cell entrapment in the lungs, for instances by decreasing MSC size, and enhancing organ-specific capture by modulating cell surface properties, it could be expected an improvement in the target efficiency of MSC-based therapies [181, 185].

Overall, the recent advances in genetic engineering protocols for MSCs and clinical studies assessing MSC-based therapy products herein described will certainly contribute to increase our knowledge on the safety and therapeutic potential of MSCs, instructing future trials to further develop this rapidly expanding field. However, further studies are required before therapies with genetically engineered MSCs can be routinely used in a clinical setting.

# REFERENCES

[1]    L. Xie, Zhang, N. Marsano, A. Vunjak-Novakovic, G., Zhang, Y., and Lopez, M. J., "In vitro mesenchymal trilineage differentiation and extracellular matrix production by adipose and bone marrow derived adult equine multipotent stromal cells on a collagen scaffold," *Stem Cell Rev. Reports*, vol. 9, no. 6, pp. 858–872, 2013.

[2]    Hofer, H. R. and Tuan, R. S., "Secreted trophic factors of mesenchymal stem cells support neurovascular and musculoskeletal therapies," *Stem Cell Res. Ther.*, vol. 7, no. 1, pp. 1–14, 2016.

[3]    F. Gao et al., "Mesenchymal stem cells and immunomodulation: current status and future prospects," *Cell Death Dis.*, vol. 7, p. e2062, 2016.

[4]    I. K. and Hiwatashi., R. C. B. Nao, Bing, Renjie, "Mesenchymal stem cells have anti-fibrotic effects on Transforming Growth Factor-β1-stimulated vocal fold fibroblasts," *Laryngoscope*, vol. 127, pp. 35–41, 2017.

[5]    He, A., Jiang, Y., Gui, C., Sun, Y., Li, J., and Wang, J. A., "The antiapoptotic effect of mesenchymal stem cell transplantation on ischemic myocardium is enhanced by anoxic preconditioning," *Can. J. Cardiol.*, vol. 25, no. 6, pp. 353–358, 2009.

[6]    Murray, I. R., and Péault, B., "Q&A: Mesenchymal stem cells - where do they come from and is it important?," *BMC Biol.*, vol. 13, no. 1, pp. 4–9, 2015.

[7] Friedenstein, A. F., Chailakhjan, R. H., and Lalykina, K. S., "The development of fibroblast colonies in marrow and spleen cells monolayer cultures of guinea-pig bone," *Development*, vol. 38, no. 4, pp. 1–6, 1970.

[8] Yu, A. Y. F. Gorskaya, F., and Kulagina, N. N., "Precursor cells of fibroblasts detected by in vitro cloning of cells from hematopoietic organs of normal and irradiated mice," *Bull. Exp. Biol. Med.*, vol. 81, no. 5, pp. 195–198, 1976.

[9] Caplan, Arnold I., "Mesenchymal Stem Cells," *J. Orthop. Res.*, no. 9, pp. 641–650, 1991.

[10] Cohnheim, J., "Ueber Entzündung und Eiterung, [About inflammation and suppuration]" *Archiv für pathologische Anatomie und Physiologie und für klinische Medicin* [Archive for pathological anatomy and physiology and for clinical medicine]. Mendeley Ltd., London, pp. 1–79, 1867.

[11] Hernigou, P., "Bone transplantation and tissue engineering, part IV. Mesenchymal stem cells: history in orthopedic surgery from Cohnheim and Goujon to the Nobel Prize of Yamanaka," *Int. Orthop.*, vol. 39, no. 4, pp. 807–817, 2015.

[12] Dominici, M. et al., "Minimal criteria for defining multipotent mesenchymal stromal cells. The International Society for Cellular Therapy position statement.," *Cytotherapy*, vol. 8, no. 4, pp. 315–7, 2006.

[13] Murray, I. R. et al., "Natural history of mesenchymal stem cells, from vessel walls to culture vessels," *Cell. Mol. Life Sci.*, vol. 71, no. 8, pp. 1353–1374, 2014.

[14] Pontikoglou, C., F. Deschaseaux, Sensebé, L., and Papadaki, H. A., "Bone Marrow Mesenchymal Stem Cells: Biological Properties and Their Role in Hematopoiesis and Hematopoietic Stem Cell Transplantation," *Stem Cell Rev. Reports*, vol. 7, no. 3, pp. 569–589, 2011.

[15] Nagamura-Inoue, T., "Umbilical cord-derived mesenchymal stem cells: Their advantages and potential clinical utility," *World J. Stem Cells*, vol. 6, no. 2, p. 195, 2014.

[16]  Kim, D. W., Staples, M., Shinozuka, K., Pantcheva, P., Kang, S. D., and Borlongan, C. V., "Wharton's jelly-derived mesenchymal stem cells: Phenotypic characterization and Optimizing their therapeutic potential for clinical applications," *Int. J. Mol. Sci.*, vol. 14, no. 6, pp. 11692–11712, 2013.

[17]  Abumaree, M. H. et al., "Phenotypic and Functional Characterization of Mesenchymal Stem Cells from Chorionic Villi of Human Term Placenta," *Stem Cell Rev. Reports*, vol. 9, no. 1, pp. 16–31, 2013.

[18]  Uzieliene, I., Urbonaite, G., Tachtamisevaite, Z., Mobasheri, A., and Bernotiene, E., "The Potential of Menstrual Blood-Derived Mesenchymal Stem Cells for Cartilage Repair and Regeneration: Novel Aspects," *Stem Cells Int.*, vol. 2018, pp. 1–10, 2018.

[19]  K. G. M. and Minteer, J. P. R. Danielle, "Adipose-Derived Mesenchymal Stem Cells: Biology and Potential Applications," *Adv. Biochem. Eng. 59–71*, no. 2012, pp. 59–70, 2012.

[20]  Fawzy El-Sayed, K. M. and Dörfer, C. E., "Gingival Mesenchymal Stem/Progenitor Cells: A Unique Tissue Engineering Gem," *Stem Cells Int.*, vol. 2016, pp. 1–16, 2016.

[21]  L. M. E., M. N. V. M., and S. O. E., "Mesenchymal stem cells derived from dental pulp: A review," *Stem Cells Int.*, vol. 2016, 2016.

[22]  Elahi, K. C., Klein, G., Avci-Adali, M., Sievert, K. D., Macneil, S., and Aicher, W. K., "Human mesenchymal stromal cells from different sources diverge in their expression of cell surface proteins and display distinct differentiation patterns," *Stem Cells Int.*, vol. 2016, no. Figure 1, 2016.

[23]  Hass, R., Kasper, C., Böhm, S., and Jacobs, R., "Different populations and sources of human mesenchymal stem cells (MSC): A comparison of adult and neonatal tissue-derived MSC," *Cell Commun. Signal.*, vol. 9, no. 1, p. 12, 2011.

[24]  Rebelatto, C. K., et al., "Dissimilar Differentiation of Mesenchymal Stem Cells from Bone Marrow, Umbilical Cord Blood, and Adipose Tissue," *Exp. Biol. Med.*, vol. 233, no. 7, pp. 901–913, 2008.

[25] Kern, S., Eichler, H., Stoeve, J., Klüter, H., and Bieback, K., "Comparative Analysis of Mesenchymal Stem Cells from Bone Marrow, Umbilical Cord Blood, or Adipose Tissue," *Stem Cells*, vol. 24, no. 5, pp. 1294–1301, 2006.

[26] Sarvar, D. Shamsasenjan, P. K., and Akbarzadehlaleh, P., "Mesenchymal stem cell-derived exosomes: New opportunity in cell-free therapy," *Adv. Pharm. Bull.*, vol. 6, no. 3, pp. 293–299, 2016.

[27] Harrell, C. R., et al., "Therapeutic potential of mesenchymal stem cell-derived exosomes in the treatment of eye diseases," *Adv. Exp. Med. Biol.*, vol. 1089, pp. 47–57, 2018.

[28] Han, C., et al., "Human umbilical cord mesenchymal stem cell derived exosomes encapsulated in functional peptide hydrogels promote cardiac repair," *Biomater. Sci.*, 2019.

[29] Shi, Y., Shi, H., Nomi, A., and Lei-lei, Z., "Mesenchymal stem cell–derived extracellular vesicles: a new impetus of promoting angiogenesis in tissue regeneration," *Cytotherapy*, vol. 21, no. 5, pp. 497–508, 2019.

[30] Chulpanova, D. S., Kitaeva, K. V., Tazetdinova, L. G., James, V., Rizvanov, A. A., and Solovyeva, V. V., "Application of Mesenchymal stem cells for therapeutic agent delivery in anti-tumor treatment," *Front. Pharmacol.*, vol. 9, no. MAR, pp. 1–10, 2018.

[31] A. O. S. and Buitenhuis, M., "Molecular mechanisms underlying adhesion and migration of hematopoietic stem cells," *Cell Adh. Migr.*, no. October, pp. 39–48, 2012.

[32] Hsuan, Y. C., Lin, C., Chang, C., and Lin, M., "Mesenchymal stem cell-based treatments for stroke, neural trauma, and heat stroke," *Brain Behav.*, vol. 6, no. 10, pp. 1–11, 2016.

[33] Zhao, R. C., *Essentials of mesenchymal stem cell biology and its clinical translation*. 2013.

[34] Zachar, L., Bačenková, D., and Rosocha, J., "Activation, homing, and role of the mesenchymal stem cells in the inflammatory environment," *J. Inflamm. Res.*, vol. 9, pp. 231–240, 2016.

[35] Yin, F., Battiwalla, M., Ito, S., Feng, X., and Chinian, F., "Bone marrow mesenchymal stromal cells to treat tissue damage in allogeneic stem cell transplant recipients," *Stem Cell*, vol. 32, no. 5, pp. 1278–1288, 2014.

[36] Katuchova, J. et al., "Impact of different pancreatic micro-environments on improvement in hyperglycemia and insulin deficiency in diabetic rats after transplantation of allogeneic mesenchymal stromal cells," *J. Surg. Res.*, vol. 178, no. 1, pp. 188–195, 2012.

[37] Glenn, J. D., M. D. Smith, P. A. Calabresi, and K. A. Whartenby, "Mesenchymal stem cells differentially modulate effector CD8+ T cell subsets and exacerbate experimental autoimmune encephalo-myelitis," *Stem Cells*, vol. 32, no. 10, pp. 2744–2755, 2014.

[38] D. Cizkova et al., "Repetitive Intrathecal Catheter Delivery of Bone Marrow Mesenchymal Stromal Cells Improves Functional Recovery in a Rat Model of Contusive Spinal Cord Injury," *J. Neurotrauma*, vol. 28, no. 9, pp. 1951–1961, 2010.

[39] Ankrum, J. A., Ong, J. F., and Karp, J. M., "Mesenchymal stem cells: immune evasive, not immune privileged," vol. 32, no. 3, pp. 252–260, 2014.

[40] Tse, W. T., Pendleton, J. D., Beyer, W. M., Egalka, M. C., and Guinan E. C., "Suppression of allogeneic T-cell proliferation by human marrow stromal cells: implications in transplantation.," *Trans-plantation*, vol. 75, no. 3, pp. 389–97, 2003.

[41] Sharma R. R., Pollock., K., Hubel, A., and McKenna, D., "Mesenchymal stem or stromal cells: a review of clinical applications and manufacturing practices," *Transfusion*, vol. 5, no. 6, pp. 1418–1437, 2016.

[42] Undale, A. H., Westendorf, J. J., Yaszemski, M. J., and Khosla, S., "Mesenchymal stem cells for bone repair and metabolic bone diseases.," *Mayo Clin. Proc.*, vol. 84, no. 10, pp. 893–902, 2009.

[43] Orlic D., et al., "Mobilized bone marrow cells repair the infarcted heart, improving function and survival," *Proc. Natl. Acad. Sci.*, vol. 98, no. 18, pp. 10344–10349, 2002.

[44] Freyman, T. et al., "A quantitative, randomized study evaluating three methods of mesenchymal stem cell delivery following myocardial infarction," *Eur. Heart J.*, vol. 27, no. 9, pp. 1114–1122, 2006.

[45] Zhang W.-W., et al., "The First Approved Gene Therapy Product for Cancer Ad- *p53* (Gendicine): 12 Years in the Clinic," *Hum. Gene Ther.*, vol. 29, no. 2, pp. 160–179, 2018.

[46] Salmon, F., Grosios, K., and Petry, H., "Safety profile of recombinant adeno-associated viral vectors: Focus on alipogene tiparvovec (Glybera®)," *Expert Rev. Clin. Pharmacol.*, vol. 7, no. 1, pp. 53–65, 2014.

[47] Mosca, J. D. et al., "Mesenchymal stem cells as vehicles for gene delivery [In Process Citation]," *Clin Orthop*, no. 379 Suppl, pp. S71-90, 2000.

[48] Kenneth Lundstrom PanTherapeutics, "Viral vectors in gene therapy," *Period. Biol.*, vol. 103, no. 2, pp. 139–143, 2018.

[49] A. Telesnitsky, "Retroviruses: Molecular Biology, Genomics and Pathogenesis," *Future Virol.*, vol. 5, no. 5, pp. 539–543, 2010.

[50] H. F. and Maeda, Y. Y. Naoyoshi, "Oncogenesis by retroviruses: old and new paradigms," *Rev. Med. Virol.*, vol. 7, no. 9, pp. 514–522, 2008.

[51] Miguel, D. De, Lemke, J., Anel, A., Walczak, H., and Martinez-Lostao, L., "Onto better TRAILs for cancer treatment," *Cell Death Differ.*, vol. 23, no. 5, pp. 733–747, 2016.

[52] Grisendi G., et al., "Adipose-derived mesenchymal stem cells as stable source of tumor necrosis factor-related apoptosis-inducing ligand delivery for cancer therapy," *Cancer Res.*, vol. 70, no. 9, pp. 3718–3729, 2010.

[53] Mounayar M., Magee, C. N., and Abdi, R., "Immunomodulation by mesenchymal stem cells - a potential therapeutic strategy for type 1 diabetes," *Stem Cell-Dependent Ther. Mesenchymal Stem Cells Chronic Inflamm. Disord.*, vol. 57, no. July, pp. 309–318, 2013.

[54] Fierro, F. A., Sierralta, W. D., Epuñan, M. J., and Minguell, J. J., "Marrow-derived mesenchymal stem cells: Role in epithelial tumor

cell determination," *Clin. Exp. Metastasis*, vol. 21, no. 4, pp. 313–319, 2004.

[55] W. Hu et al., "Human umbilical blood mononuclear cell-derived mesenchymal stem cells serve as interleukin-21 gene delivery vehicles for epithelial ovarian cancer therapy in nude mice," *Biotechnol. Appl. Biochem.*, vol. 58, no. 6, pp. 397–404, 2011.

[56] Elzaouk L., Moelling K., and Pavlovic J., "Anti-tumor activity of mesenchymal stem cells producing IL-12 in a mouse melanoma model," *Exp. Dermatol.*, vol. 15, no. 11, pp. 865–874, 2006.

[57] Ursula A., et al., "Prodrug suicide gene therapy for cancer targeted intracellular by mesenchymal stem cell exosomes," *Int. J. Cancer*, vol. 144, no. 4, pp. 897–908, 2019.

[58] Kalimuthu S. et al., "Genetically engineered suicide gene in mesenchymal stem cells using a Tet-On system for anaplastic thyroid cancer," *PLoS One*, vol. 12, no. 7, pp. 1–19, 2017.

[59] Miao, C., Lei, M., Hu, W., Han, S., andWang, Q., "A brief review: The therapeutic potential of bone marrow mesenchymal stem cells in myocardial infarction," *Stem Cell Res. Ther.*, vol. 8, no. 1, pp. 4–9, 2017.

[60] Leibacher J., and Henschler, R., "Biodistribution, migration and homing of systemically applied mesenchymal stem/stromal cells Mesenchymal Stem/Stromal Cells - An update," *Stem Cell Res. Ther.*, vol. 7, no. 1, pp. 1–12, 2016.

[61] Cheng Z. et al., "Targeted migration of mesenchymal stem cells modified with CXCR4 gene to infarcted myocardium improves cardiac performance," *Mol. Ther.*, vol. 16, no. 3, pp. 571–579, 2008.

[62] Fiarresga A. et al., "Intracoronary delivery of human mesenchymal/stromal stem cells: Insights from coronary micro-circulation invasive assessment in a swine model," *PLoS One*, vol. 10, no. 10, pp. 1–12, 2015.

[63] Yaniz-Galende E. and Hajjar R. J., *Stem cell and gene therapy for cardiac regeneration*, vol. 1. Woodhead Publishing Limited, 2014.

[64] Naldini L. et al., "In vivo gene delivery and stable transduction of nondividing cells by a lentiviral vector," *Science (80-.).*, vol. 272, no. 5259, pp. 263–267, 1996.

[65] Bartholomae, C. C. et al., "The genotoxic potential of retroviral vectors is strongly modulated by vector design and integration site selection in a mouse model of HSC gene therapy," *J. Clin. Invest.*, vol. 119, no. 4, pp. 964–975, 2009.

[66] Jiang, W. et al., "An optimized method for high-titer lentivirus preparations without ultracentrifugation," *Sci. Rep.*, vol. 5, pp. 1–9, 2015.

[67] Sena-Esteves, Tebbets, M. J. C., Steffens S., Crombleholme T., and Flake, A. W., "Optimized large-scale production of high titer lentivirus vector pseudotypes," *J. Virol. Methods*, vol. 122, no. 2, pp. 131–139, 2004.

[68] Chen X., Wang K., Chen S., and Chen Y., "Effects of mesenchymal stem cells harboring the Interferon-β gene on A549 lung cancer in nude mice," *Pathol. Res. Pract.*, vol. 215, no. 3, pp. 586–593, 2019.

[69] Ahn, J. ok, Lee, H. woo, Seo, K. won, Kang, S. keun, Ra, J. chan , and Youn, H. young, "Anti-Tumor Effect of Adipose Tissue Derived-Mesenchymal Stem Cells Expressing Interferon-β and Treatment with Cisplatin in a Xenograft Mouse Model for Canine Melanoma," *PLoS One*, vol. 8, no. 9, pp. 1–11, 2013.

[70] Dembinski J. L. et al., "Tumor stroma engraftment of gene-modified mesenchymal stem cells as anti-tumor therapy against ovarian cancer," *Cytotherapy*, vol. 15, no. 1, pp. 20-32.e2, 2013.

[71] Zhu X. et al., "Gene therapy of gastric cancer using LIGHT-secreting human umbilical cord blood-derived mesenchymal stem cells," *Gastric Cancer*, vol. 16, no. 2, pp. 155–166, 2013.

[72] Mao W. et al., "TNF-α expression in the UCB-MSCs as stable source inhibits gastric cancers growth in nude mice," *Cancer Invest.*, vol. 30, no. 6, pp. 463–472, 2012.

[73] Boucher J. et al., "Apelin, a newly identified adipokine up-regulated by insulin and obesity," *Endocrinology*, vol. 146, no. 4, pp. 1764–1771, 2005.

[74] Japp A. G. et al., "Acute cardiovascular effects of apelin in humans: Potential role in patients with chronic heart failure," *Circulation*, vol. 121, no. 16, pp. 1818–1827, 2010.

[75] Gao L. R. et al., "Overexpression of apelin in Wharton' jelly mesenchymal stem cell reverses insulin resistance and promotes pancreatic β cell proliferation in type 2 diabetic rats," *Stem Cell Res. Ther.*, vol. 10, no. 1, pp. 1–14, 2019.

[76] Bougioukli S. et al., "Lentiviral Gene Therapy For Bone Repair Using Human Umbilical Cord Blood Derived-Mesenchymal Stem Cells," *Hum. Gene Ther.*, no. 323, pp. 1–32, 2019.

[77] Kim,Y., Kang, B. J., Kim, W. H., Yun, H. S., and Kweon, O. K., "Evaluation of mesenchymal stem cell sheets overexpressing BMP-7 in canine critical-sized bone defects," *Int. J. Mol. Sci.*, vol. 19, no. 7, 2018.

[78] Zhang F. et al., "Role of FGF-2 Transfected Bone Marrow Mesenchymal Stem Cells in Engineered Bone Tissue for Repair of Avascular Necrosis of Femoral Head in Rabbits," *Cell. Physiol. Biochem.*, vol. 48, no. 2, pp. 773–784, 2018.

[79] Loebinger, M. R., Eddaoudi A., Davies D., and Janes S. M., "Mesenchymal stem cell delivery of TRAIL can eliminate metastatic cancer," *Cancer Res.*, vol. 69, no. 10, pp. 4134–4142, 2009.

[80] Deng Q. et al., "TRAIL-secreting mesenchymal stem cells promote apoptosis in heat-shock-treated liver cancer cells and inhibit tumor growth in nude mice," *Br. Dent. J.*, vol. 217, no. 1, pp. 317–327, 2014.

[81] Yan C. et al., "Suppression of orthotopically implanted hepatocarcinoma in mice by umbilical cord-derived mesenchymal stem cells with sTRAIL gene expression driven by AFP promoter," *Biomaterials*, vol. 35, no. 9, pp. 3035–3043, 2014.

[82] Spano C. et al., "Soluble TRAIL Armed Human MSC As Gene Therapy For Pancreatic Cancer," *Sci. Rep.*, vol. 9, no. 1, pp. 1–14, 2019.

[83] Rossignoli F. et al., "MSC-delivered soluble TRAIl and paclitaxel as novel combinatory treatment for pancreatic adenocarcinoma," *Theranostics*, vol. 9, no. 2, pp. 436–448, 2019.

[84] Xia L. et al., "TRAIL-expressing gingival-derived mesenchymal stem cells inhibit tumorigenesis of tongue squamous cell carcinoma," *J. Dent. Res.*, vol. 94, no. 1, pp. 219–228, 2015.

[85] Luetzkendorf, J. Mueller, L. P., Mueller, T., Caysa, H., Nerger, K., and Schmoll, H. J., "Growth inhibition of colorectal carcinoma by lentiviral TRAIL-transgenic human mesenchymal stem cells requires their substantial intratumoral presence," *J. Cell. Mol. Med.*, vol. 14, no. 9, pp. 2292–2304, 2010.

[86] Y. C. et al., "Human umbilical cord mesenchymal stem cells as vehicles of CD20-specific TRAIL fusion protein delivery: A double-target therapy against non-Hodgkin's lymphoma," *Mol. Pharm.*, vol. 10, no. 1, pp. 142–151, 2013.

[87] Liu Z. et al., "Lentivirus-mediated microRNA-26a overexpression in bone mesenchymal stem cells facilitates bone regeneration in bone defects of calvaria in mice," *Mol. Med. Rep.*, vol. 18, no. 6, pp. 5317–5326, 2018.

[88] Lang, F. M. et al., "Mesenchymal stem cells as natural biofactories for exosomes carrying miR-124a in the treatment of gliomas," *Neuro. Oncol.*, vol. 20, no. 3, pp. 380–390, 2018.

[89] Chen, C., Li H., Jiang J., Zhang Q., and Yan F., "Inhibiting PHD2 in bone marrow mesenchymal stem cells via lentiviral vector-mediated RNA interference facilitates the repair of periodontal tissue defects in SD rats," *Oncotarget*, vol. 8, no. 42, pp. 72676–72699, 2017.

[90] Lin P., Correa, D., Lin, Y., and Caplan, A. I., "Polybrene inhibits human mesenchymal stem cell proliferation during lentiviral transduction," *PLoS One*, vol. 6, no. 8, 2011.

[91] Deryabin, P., Griukova, A., Shatrova, A., Petukhov, A., Nikolsky, N., and Borodkina, A., "Optimization of lentiviral transduction parameters and its application for CRISPR-based secretome modification of human endometrial mesenchymal stem cells," *Cell Cycle*, vol. 18, no. 6–7, pp. 742–758, 2019.

[92]  Wold, W. S. M. and Toth K., "Adenovirus vectors for gene therapy, vaccination and cancer gene therapy.," *Curr. Gene Ther.*, vol. 13, no. 6, pp. 421–33, 2013.

[93]  Hartigan-O'connor D., Barjot C., Salvatori G., and Chamberlain J. S., "Generation a n d Growth of Gutted Adenoviral Vectors," in *Methods in Enzymology*, vol. 346, 2002, pp. 224–246.

[94]  Ackermann, D. M. et al., *List of Contributors of Volume 1*, vol. 2, no. Volume 2. 2018.

[95]  Kaspar, B. K., "Gene therapy: Direct viral delivery," *Encycl. Neurosci.*, pp. 633–639, 2010.

[96]  Marasini S. et al., "Effects of Adenoviral Gene Transduction on the Stemness of Human Bone Marrow Mesenchymal Stem Cells," *Mol. Cells*, vol. 40, no. 8, pp. 598–605, 2017.

[97]  Kanehira M. et al., "Targeted delivery of NK4 to multiple lung tumors by bone marrow-derived mesenchymal stem cells," *Cancer Gene Ther.*, vol. 14, no. 11, pp. 894–903, 2007.

[98]  Qiao B., Shui W., Cai L., Guo S., and Jiang D., "Human mesenchymal stem cells as delivery of osteoprotegerin gene: Homing and therapeutic effect for osteosarcoma," *Drug Des. Devel. Ther.*, vol. 9, pp. 969–976, 2015.

[99]  Chen Q. et al., "Therapeutic potential of bone marrow-derived mesenchymal stem cells producing pigment epithelium-derived factor in lung carcinoma," *Int. J. Mol. Med.*, vol. 30, no. 3, pp. 527–534, 2012.

[100] Castleton A. et al., "Human mesenchymal stromal cells deliver systemic oncolytic measles virus to treat acute lymphoblastic leukemia in the presence of humoral immunity," *Blood*, vol. 123, no. 9, pp. 1327–1335, 2014.

[101] Kaczorowski A. et al., "Delivery of improved oncolytic adenoviruses by mesenchymal stromal cells for elimination of tumorigenic pancreatic cancer cells," vol. 7, no. 8, 2016.

[102] Prieto-Vila M., Takahashi R. U., Usuba W., Kohama I., and Ochiya T., "Drug resistance driven by cancer stem cells and their niche," *Int. J. Mol. Sci.*, vol. 18, no. 12, 2017.

[103] Farrell E. et al., "vIL-10-overexpressing human MSCs modulate naïve and activated T lymphocytes following induction of collagenase-induced osteoarthritis," *Stem Cell Res. Ther.*, vol. 7, no. 1, pp. 1–11, 2016.

[104] Tian S., Yan Y., Qi X., Li X., and Li Z., "Treatment of Type II Collagen-Induced Rat Rheumatoid Arthritis Model by Interleukin 10 (IL10)-Mesenchymal Stem Cells (BMSCs)," *Med. Sci. Monit.*, vol. 25, pp. 2923–2934, 2019.

[105] Yin T. et al., "Genetically modified human placenta-derived mesenchymal stem cells with FGF-2 and PDGF-BB enhance neovascularization in a model of hindlimb ischemia," *Mol. Med. Rep.*, vol. 12, no. 4, pp. 5093–5099, 2015.

[106] Rennert R. C. et al., "Diabetes impairs the angiogenic potential of adipose-derived stem cells by selectively depleting cellular subpopulations," *Stem Cell Res. Ther.*, vol. 5, no. 3, pp. 1–12, 2014.

[107] Cronk, S. M. et al., "Adipose-Derived Stem Cells From Diabetic Mice Show Impaired Vascular Stabilization in a Murine Model of Diabetic Retinopathy," *Stem Cells Transl. Med.*, vol. 4, no. 5, pp. 459–467, 2015.

[108] Peng Z. et al., "Glyoxalase-1 Overexpression Reverses Defective Proangiogenic Function of Diabetic Adipose-Derived Stem Cells in Streptozotocin-Induced Diabetic Mice Model of Critical Limb Ischemia," *Stem Cells Transl. Med.*, vol. 6, no. 1, pp. 261–271, 2017.

[109] Kim H. K. et al., "Alterations in the Proangiogenic Functions of Adipose Tissue–Derived Stromal Cells Isolated from Diabetic Rats," *Stem Cells Dev.*, vol. 17, no. 4, pp. 669–680, 2008.

[110] Zhang X. et al., "Netrin-1 improves adipose-derived stem cell proliferation, migration, and treatment effect in type 2 diabetic mice with sciatic denervation 11 Medical and Health Sciences 1103 Clinical Sciences," *Stem Cell Res. Ther.*, vol. 9, no. 1, pp. 1–13, 2018.

[111] Abreu, C. E. C. V., Ferreira P. P. R., Moraes, F. Y. de, Neves Jr., W. F. P., Gadia, R., and Carvalho, H. de A., "Stereotactic body

radiotherapy in lung cancer: an update," *J. Bras. Pneumol.*, vol. 41, no. 4, pp. 376–387, 2015.

[112] Liu D. et al., "Decorin-Modified Umbilical Cord Mesenchymal Stem Cells (MSCs) Attenuate Radiation-Induced Lung Injuries via Regulating Inflammation, Fibrotic Factors, and Immune Responses," *Int. J. Radiat. Oncol. Biol. Phys.*, vol. 101, no. 4, pp. 945–956, 2018.

[113] Hu, M. C. et al., "Klotho: a novel phosphaturic substance acting as an autocrine enzyme in the renal proximal tubule," *FASEB J.*, vol. 24, no. 9, pp. 3438–3450, 2010.

[114] Zhang F., Wan X., Cao Y. Z., Sun D., and Cao C. C., "Klotho gene-modified BMSCs showed elevated antifibrotic effects by inhibiting the Wnt/β-catenin pathway in kidneys after acute injury," *Cell Biol. Int.*, vol. 42, no. 12, pp. 1670–1679, 2018.

[115] L. Orozco et al., "Treatment of knee osteoarthritis with autologous mesenchymal stem cells: A pilot study," *Transplantation*, vol. 95, no. 12, pp. 1535–1541, 2013.

[116] J. Do Kim et al., "Clinical outcome of autologous bone marrow aspirates concentrate (BMAC) injection in degenerative arthritis of the knee," *Eur. J. Orthop. Surg. Traumatol.*, vol. 24, no. 8, pp. 1505–1511, 2014.

[117] Gigante A., Cecconi S., Calcagno S., Busilacchi A., and Enea, D., "Arthroscopic Knee Cartilage Repair With Covered Microfracture and Bone Marrow Concentrate," *Arthrosc. Tech.*, vol. 1, no. 2, pp. e175–e180, 2012.

[118] Welton K. L., Logterman S., Bartley J. H., Vidal A. F., and McCarty E. C., "Knee Cartilage Repair and Restoration: Common Problems and Solutions," *Clin. Sports Med.*, vol. 37, no. 2, pp. 307–330, 2018.

[119] Yang S., Qian Z., Liu D., Wen N., Xu J., and Guo X., "Integration of C-type natriuretic peptide gene-modified bone marrow mesenchymal stem cells with chitosan/silk fibroin scaffolds as a promising strategy for articular cartilage regeneration," *Cell Tissue Bank.*, vol. 0123456789, 2019.

[120] Xiao W., Gao G., Ling C., Herzog R. W., Xiao X., and Samulski R. J., "Impact of neutralizing antibodies against AAV is a key

consideration in gene transfer to nonhuman primates," *Nat. Med.*, vol. 24, no. 6, pp. 699–699, 2018.

[121] Strobel B. et al., "Standardized, Scalable, and Timely Flexible Adeno-Associated Virus Vector Production Using Frozen High-Density HEK-293 Cell Stocks and CELLdiscs," *Hum. Gene Ther. Methods*, vol. 30, no. 1, pp. 23–33, 2019.

[122] Ren C., Kumar, S., Chanda, D., Chen, J., Mountz, J. D., and Ponnazhagan, S., "Therapeutic potential of mesenchymal stem cells producing IFN- α in a mouse melanoma lung metastasis model," vol. 26, no. 9, pp. 2332–2338, 2008.

[123] Jin, S. et al., "Mesenchymal Stem Cells with Enhanced Bcl-2 Expression Promote Liver Recovery in a Rat Model of Hepatic Cirrhosis," *Cell. Physiol. Biochem.*, vol. 40, no. 5, pp. 1117–1128, 2016.

[124] Liao, H., Zhong Z., Liu Z., Li L., Ling Z., and Zou X., "Bone mesenchymal stem cells co-expressing VEGF and BMP-6 genes to combat avascular necrosis of the femoral head," *Exp. Ther. Med.*, vol. 15, no. 1, pp. 954–962, 2018.

[125] Nakajima M. et al., "Mesenchymal Stem Cells Overexpressing Interleukin-10 Promote Neuroprotection in Experimental Acute Ischemic Stroke," *Mol. Ther. - Methods Clin. Dev.*, vol. 6, no. September, pp. 102–111, 2017.

[126] Salazar, V. S., Gamer, L. W., and Rosen, V., "BMP signalling in skeletal development, disease and repair," *Nat. Rev. Endocrinol.*, vol. 12, no. 4, pp. 203–221, 2016.

[127] Kumar, S., Nagy, T. R., and Ponnazhagan, S., "Therapeutic potential of genetically modified adult stem cells for osteopenia," *Gene Ther.*, vol. 17, no. 1, pp. 105–116, 2010.

[128] Bao C., Guo, J., Lin, G., Hu, M., and Hu, Z., "TNFR gene-modified mesenchymal stem cells attenuate inflammation and cardiac dysfunction following MI," *Scand. Cardiovasc. J.*, vol. 42, no. 1, pp. 56–62, 2008.

[129] Gao, X. et al., "Protective effects of mesenchymal stem cells overexpressing extracellular regulating kinase 1/2 against stroke in rats," *Brain Res. Bull.*, vol. 149, pp. 42–52, 2019.

[130] Yin H., Kanasty, R. L., Eltoukhy, A. A., A. Vegas, J., Dorkin, J. R., and Anderson D. G., "Non-viral vectors for gene-based therapy," *Nat. Rev. Genet.*, vol. 15, no. 8, pp. 541–555, Aug. 2014.

[131] Yin H., Kanasty R. L., Eltoukhy, A. A., Vegas, A. J., Dorkin, J. R., and Anderson, D. G., "Non-viral vectors for gene-based therapy," *Nat. Rev. Genet.*, vol. 15, no. 8, pp. 541–555, Aug. 2014.

[132] Ramamoorth M. and Narvekar A., "Non viral vectors in gene therapy- an overview.," *J. Clin. Diagn. Res.*, vol. 9, no. 1, pp. 1–6, Jan. 2015.

[133] Santos, J. L., Pandita D., Rodrigues J., Pego A. P., Granja P. L., and Tomas H., "Non-Viral Gene Delivery to Mesenchymal Stem Cells: Methods, Strategies and Application in Bone Tissue Engineering and Regeneration," *Curr. Gene Ther.*, vol. 11, no. 1, pp. 46–57, 2011.

[134] Hamann A., Nguyen A., and Pannier A. K., "Nucleic acid delivery to mesenchymal stem cells: A review of nonviral methods and applications," *J. Biol. Eng.*, vol. 13, no. 1, pp. 1–16, 2019.

[135] Xiang S. et al., "Uptake mechanisms of non-viral gene delivery," *J. Control. Release*, vol. 158, no. 3, pp. 371–378, Mar. 2012.

[136] Cho J. W., C. Lee Y., and Ko Y., "Therapeutic potential of mesenchymal stem cells overexpressing human forkhead box A2 gene in the regeneration of damaged liver tissues," *J. Gastroenterol. Hepatol.*, vol. 27, no. 8, pp. 1362–1370, 2012.

[137] Tsubokawa T. et al., "Impact of anti-apoptotic and anti-oxidative effects of bone marrow mesenchymal stem cells with transient overexpression of heme oxygenase-1 on myocardial ischemia," *Am. J. Physiol. Circ. Physiol.*, vol. 298, no. 5, pp. 1320–1329, 2010.

[138] You M. H. et al., "Cytosine deaminase-producing human mesenchymal stem cells mediate an antitumor effect in a mouse xenograft model," *J. Gastroenterol. Hepatol.*, vol. 24, no. 8, pp. 1393–1400, 2009.

[139] Yu Y. et al., "Knockdown of MicroRNA Let-7a Improves the Functionality of Bone Marrow-Derived Mesenchymal Stem Cells in Immunotherapy," *Mol. Ther.*, vol. 25, no. 2, pp. 480–493, 2017.

[140] Teoh H. K. et al., "Small interfering RNA silencing of interleukin-6 in mesenchymal stromal cells inhibits multiple myeloma cell growth," *Leuk. Res.*, vol. 40, pp. 44–53, 2015.

[141] Li, W. et al., "Bcl-2 Engineered MSCs Inhibited Apoptosis and Improved Heart Function," *Stem Cells*, vol. 25, no. 8, pp. 2118–2127, 2007.

[142] Rejman J., Tavernier G., Bavarsad Demeester N., J., and De Smedt S. C., "mRNA transfection of cervical carcinoma and mesenchymal stem cells mediated by cationic carriers," *J. Control. Release*, vol. 147, no. 3, pp. 385–391, 2010.

[143] Park, J. S. et al., "Chondrogenesis of human mesenchymal stem cells mediated by the combination of SOX trio SOX5, 6, and 9 genes complexed with PEI-modified PLGA nanoparticles," *Biomaterials*, vol. 32, no. 14, pp. 3679–3688, 2011.

[144] Yang F. et al., "Genetic engineering of human stem cells for enhanced angiogenesis using biodegradable polymeric nano-particles," *Proc. Natl. Acad. Sci.*, vol. 107, no. 8, pp. 3317–3322, 2009.

[145] Malik Y. S. et al., "Polylysine-modified polyethylenimine polymer can generate genetically engineered mesenchymal stem cells for combinational suicidal gene therapy in glioblastoma," *Acta Biomater.*, vol. 80, pp. 144–153, 2018.

[146] Zhang,T. Y., Huang B., Yuan, Z. Y., Hu Y. L., Tabata, Y., and Gao J. Q., "Gene recombinant bone marrow mesenchymal stem cells as a tumor-targeted suicide gene delivery vehicle in pulmonary metastasis therapy using non-viral transfection," *Nanomedicine Nanotechnology, Biol. Med.*, vol. 10, no. 1, pp. 257–267, 2014.

[147] Huang B. et al., "Peptide modified mesenchymal stem cells as targeting delivery system transfected with miR-133b for the treatment of cerebral ischemia," *Int. J. Pharm.*, vol. 531, no. 1, pp. 90–100, 2017.

[148] Kim, T. H., Kim, M., Eltohamy, M., Yun, Y. R., Jang, J. H., and Kim, H. W., "Efficacy of mesoporous silica nanoparticles in delivering BMP-2 plasmid DNA for in vitro osteogenic stimulation of mesenchymal stem cells," *J. Biomed. Mater. Res. - Part A*, vol. 101 A, no. 6, pp. 1651–1660, 2013.

[149] Zhu K. et al., "Nanoparticle-enhanced generation of gene-transfected mesenchymal stem cells for in vivo cardiac repair," *Biomaterials*, vol. 74, pp. 188–199, 2016.

[150] Das, J., Choi, Y. J., Yasuda, H., and Han, J. W., "Efficient delivery of C/EBP beta gene into human mesenchymal stem cells via polyethylenimine-coated gold nanoparticles enhances adipogenic differentiation," *Sci. Rep.*, vol. 6, pp. 1–17, 2016.

[151] Muroski, M. E., Morgan, T. J., Levenson, C. W., and Strouse,G. F., "A gold nanoparticle pentapeptide: Gene fusion to induce therapeutic gene expression in mesenchymal stem cells," *J. Am. Chem. Soc.*, vol. 136, no. 42, pp. 14763–14771, 2014.

[152] Wu, H. C., Wang, T. W., Bohn, M. C., Lin, F. H., and Spector, M., "Novel magnetic hydroxyapatite nanoparticles as non-viral vectors for the glial cell line-derived neurotrophic factor gene," *Adv. Funct. Mater.*, vol. 20, no. 1, pp. 67–77, 2010.

[153] Gonzalez-Fernandez, T. et al., "Mesenchymal stem cell fate following non-viral gene transfection strongly depends on the choice of delivery vector," *Acta Biomater.*, vol. 55, pp. 226–238, 2017.

[154] Mehier-Humbert, S. and Guy, R. H., "Physical methods for gene transfer: Improving the kinetics of gene delivery into cells," *Adv. Drug Deliv. Rev.*, vol. 57, no. 5, pp. 733–753, Apr. 2005.

[155] Kim, H. J. and Im, G. I., " Electroporation-Mediated Transfer of SOX Trio Genes (SOX-5, SOX-6, and SOX-9) to Enhance the Chondrogenesis of Mesenchymal Stem Cells," *Stem Cells Dev.*, vol. 20, no. 12, pp. 2103–2114, 2011.

[156] Lee J. S., Lee, J. M., and Im, G. Il, "Electroporation-mediated transfer of Runx2 and Osterix genes to enhance osteogenesis of adipose stem cells," *Biomaterials*, vol. 32, no. 3, pp. 760–768, 2011.

[157] Kojima, R. et al., "Designer exosomes produced by implanted cells intracerebrally deliver therapeutic cargo for Parkinson's disease treatment," *Nat. Commun.*, vol. 9, no. 1, 2018.

[158] Liu, S. P. et al., "Nonsenescent Hsp27-upregulated MSCs implantation promotes neuroplasticity in stroke model," *Cell Transplant.*, vol. 19, no. 10, pp. 1261–1279, 2010.

[159] Xu, X. et al., "Efficient homology-directed gene editing by CRISPR/Cas9 in human stem and primary cells using tube electroporation," *Sci. Rep.*, vol. 8, no. 1, pp. 1–11, 2018.

[160] Nakashima, S., Matsuyama, Y., Nitta, A., Sakai, Y., and Ishiguro, N., "Highly Efficient Transfection of Human Marrow Stromal Cells by Nucleofection," *Transplant. Proc.*, vol. 37, no. 5, pp. 2290–2292, Jun. 2005.

[161] Pham, P. Van et al., "Improved differentiation of umbilical cord blood-derived mesenchymal stem cells into insulin-producing cells by PDX-1 mRNA transfection," *Differentiation*, vol. 87, no. 5, pp. 200–208, 2014.

[162] Fakiruddin, K. S., Baharuddin, P., Lim, M. N., Fakharuzi, N. A., Yusof, N. A. N. M., and Zakaria, Z., "Nucleofection optimization and in vitro antitumourigenic effect of TRAIL-expressing human adipose-derived mesenchymal stromal cells," *Cancer Cell Int.*, vol. 14, no. 1, pp. 1–13, 2014.

[163] Kim, Y. H. et al., "Growth-inhibitory effect of neurotrophin-3-secreting adipose tissue-derived mesenchymal stem cells on the D283-MED human medulloblastoma cell line," *J. Neurooncol.*, vol. 106, no. 1, pp. 89–98, 2012.

[164] Pelled, G. et al., "BMP6-engineered MSCs induce vertebral bone repair in a pig model: A pilot study," *Stem Cells Int.*, vol. 2016, 2016.

[165] Lee, J. et al., "CRISPR/Cas9 Edited sRAGE-MSCs Protect Neuronal Death in Parkinson's Disease Model," *Int. J. Stem Cells*, vol. 12, no. 1, pp. 114–124, 2019.

[166] Lim, J. et al., "Microporation is a valuable transfection method for efficient gene delivery into human umbilical cord blood-derived

mesenchymal stem cells," *BMC Biotechnol.*, vol. 10, no. 1, p. 38, 2010.

[167] Mun, J. Y., Shin, K. K., Kwon, O., Lim, Y. T., and Oh, D. B., "Minicircle microporation-based non-viral gene delivery improved the targeting of mesenchymal stem cells to an injury site," *Biomaterials*, vol. 101, pp. 310–320, 2016.

[168] Serra, J. et al., "Engineering of human mesenchymal stem/stromal cells (MSC) with VEGF-encoding minicircles for angiogenic gene therapy," *Hum. Gene Ther.*, p. hum.2018.154, 2018.

[169] Nakashima, M., Tachibana, K., Iohara K., Ito, M., Ishikawa, M., and Akamine A., "Induction of Reparative Dentin Formation by Ultrasound-Mediated Gene Delivery of Growth/Differentiation Factor 11," *Hum. Gene Ther.*, vol. 14, no. 6, pp. 591–597, 2003.

[170] Otani K., Yamahara K., Ohnishi S., Obata H., Kitamura S., and Nagaya N., "Nonviral delivery of siRNA into mesenchymal stem cells by a combination of ultrasound and microbubbles," *J. Control. Release*, vol. 133, no. 2, pp. 146–153, 2009.

[171] Haber T., Baruch L., and Machluf M., "Ultrasound-Mediated Mesenchymal Stem Cells Transfection as a Targeted Cancer Therapy Platform," *Sci. Rep.*, vol. 7, no. January, pp. 1–13, 2017.

[172] Wang G. et al., "Enhanced Homing of CXCR-4 Modified Bone Marrow-Derived Mesenchymal Stem Cells to Acute Kidney Injury Tissues by Micro-Bubble-Mediated Ultrasound Exposure," *Ultrasound Med. Biol.*, vol. 42, no. 2, pp. 539–548, 2016.

[173] Tsulaia T. V. et al., "Glass needle-mediated microinjection of macromolecules and transgenes into primary human mesenchymal stem cells," *J. Biomed. Sci.*, vol. 10, no. 3, pp. 328–336, May 2003.

[174] Han, S. W. et al., "High-efficiency DNA injection into a single human mesenchymal stem cell using a nanoneedle and atomic force microscopy," *Nanomedicine Nanotechnology, Biol. Med.*, vol. 4, no. 3, pp. 215–225, Sep. 2008.

[175] Haber, T., Baruch, L., and Machluf, M., "Ultrasound-Mediated Mesenchymal Stem Cells Transfection as a Targeted Cancer Therapy Platform," *Sci. Rep.*, vol. 7, no. January, pp. 1–13, 2017.

[176] Einem, J. C. von et al., "Treatment of advanced gastrointestinal cancer with genetically modified autologous mesenchymal stem cells: Results from the phase 1/2 TREAT-ME-1 trial," *Int. J. Cancer*, pp. 1–10, 2019.

[177] Carroll, D., "Genome engineering with zinc-finger nucleases," *Genetics*, vol. 188, no. 4, pp. 773–782, 2011.

[178] Joung, J. Keith and Sander, J., "TALENs: a widely applicable technology for targeted genome editing," *Nat Rev Mol Cell Biol.*, vol. 22, no. 6, pp. 415–422, 2010.

[179] Ratan, M. F. H. Zubair Ahmed, Son, Young-Jin, Uddin, J.-H. K. Bhuiyan Mohammad Mahtab, Yusuf, Md. Abdullah, Zaman, Sojib Bin, and L. A. B. and Cho, J. Y., "CRISPR-Cas9: a promising genetic engineering approach in cancer research," *Ther. Adv. Vaccines*, vol. 9, no. 6, pp. 259–261, 2018.

[180] Sage E. K., Thakrar R. M., and Janes S. A. M. M., "Genetically modified mesenchymal stromal cells in cancer therapy," *Cytotherapy*, vol. 18, no. 11, pp. 1435–1445, 2016.

[181] Yin J. Q., Zhu J., and Ankrum J. A., "Manufacturing of primed mesenchymal stromal cells for therapy," *Nat. Biomed. Eng.*, vol. 3, no. 2, pp. 90–104, Feb. 2019.

[182] Zhang J. et al., "The challenges and promises of allogeneic mesenchymal stem cells for use as a cell-based therapy," *Stem Cell Res. Ther.*, vol. 6, no. 234, pp. 1–7, 2012.

[183] "Rate of lung function decline in patients with lung cancer : from virtual contact to invasive procedures," 2017, vol. 72, no. Suppl 3, pp. 222–224.

[184] Yuan Z., Lourenco S. D. S., Sage E. K., Kolluri K. K., Lowdell M. W., and Janes S. M., "Cryopreservation of human mesenchymal stromal cells expressing TRAIL for human anti-cancer therapy," *Cytotherapy*, vol. 18, no. 7, pp. 860–869, 2016.

[185] Brooks A. et al., "Concise Review: Quantitative Detection and Modeling the In Vivo Kinetics of Therapeutic Mesenchymal Stem/Stromal Cells.," *Stem Cells Transl. Med.*, vol. 7, no. 1, pp. 78–86, Jan. 2018.

In: Gene Delivery
Editor: Vanessa Zimmer

ISBN: 978-1-53616-268-4
© 2019 Nova Science Publishers, Inc.

*Chapter 2*

# NUCLEIC ACID DELIVERY SYSTEMS FOR GENE THERAPY

## *Jinhyung Lee and Yupeng Chen\*, PhD*

Department of Biomedical Engineering, University of Connecticut,
Storrs, CT, US

## ABSTRACT

Gene therapy is a major breakthrough in biotechnology nowadays. The premise for a successful gene therapy is the delivery of the therapeutic nucleic acids into target cells in a specific, effective and safe manner. With the development of biotechnology and nanotechnology, many delivery systems were developed. In this chapter, a variety of gene delivery systems (physical transfection techniques, virus-based delivery vectors, chemically engineered delivery systems and bio-inspired vehicles) will be reviewed and their strengths, shortcomings and biomedical applications will be discussed.

**Keywords:** nucleic acid delivery, RNA therapy, gene editing

---

\* Corresponding Author's Email: yupeng.chen@uconn.edu.

# INTRODUCTION

Gene therapy has been long-promised to correct a variety of human genetic diseases and defects. Despite its great potential, progress in developing clinical applications has been very slow. The problem behind is the lack of safe and efficient nucleic acid delivery systems. [1]

Currently, gene therapeutic delivery offers three different applications: Gene silencing by RNAi, transient protein expression by gene delivery and stable protein expression by genome editing. RNAi therapeutic approach is a 'loss-of-function' to treat diseases, which involves limiting or preventing protein expression in a cell. [2] SiRNA therapeutics have been actively investigated due to their high specificity, simple modification, and unlimited therapeutic targets. RNAi does not incorporate into the genome of the target cell. Thus, this allows temporal treatment of the cell, which fulfills a critical factor in regulatory and safety considerations. [3, 4]

However, nucleic acids are generally unstable outside of cells, and due to their highly anionic nature, they do not easily penetrate through cell membrane. Hence, naked siRNA is unstable in the bloodstream, and immunogenic and vulnerable to nuclease enzymes. [2] Thus, proper delivery agents are necessary to transport the siRNAs into the target cells efficiently. The delivery agents are required to be non-immunogenic and should not invoke undesirable side effects. [5] Moreover, they should avoid the off-target effect of the genes in healthy cells. For example, encapsulating the nucleic acids with delivery vehicles could deliver effectively and circumvent the off-target effect. [6]

Presently, viral vectors are very efficient carriers due to their high levels of transgene transfection efficacy. Although viral vectors have been effective in delivering siRNA molecules, biosafety, especially immunogenicity, is the major concern. [7] Non-viral vectors are developed because they are generally considered to be safer than viral vectors despite losing efficiency. Several ways have been developed to deliver short interfering RNAs; by encapsulating in synthetic vehicles such as cationic liposomes/nanoparticles or siRNAs conjugated with cell penetrating peptides or specific antibodies. [2, 8]

Although RNAi is transient and can be a safer option than permanent genome editing, gene editing can completely block the protein expression and eliminating from low-level expression for long-term. [2] Gene editing is a technique to insert, replace, or delete a specific sequence of the host genome DNA. There are four different nucleases which have been used: Meganuclease, zinc finger nucleases (ZFN), transcription activator-like effector-based nuclease (TALEN), and the clustered regularly interspaced short palindromic repeats (CRISPR)-associated protein (Cas) system. [9]

Delivering the CRISPR/Cas9 via cargoes, three approaches are commonly reported: 1) genetically encoded the Cas9 protein and the guide RNA in a plasmid, 2) mRNA for Cas9 translation with separated guide RNA, and 3) Cas9 protein with guide RNA. As a result of CRISPR/Cas9 delivery, gene gain or loss of function therapy has been developed. [9] All genome editing components are biomacromolecules that must overcome several barriers to be successfully delivered inside target cells. Viral vectors have been promising candidates because of their potential to facilitate the *in vivo* delivery of CRISPR-Cas genome editing systems. Especially, with the development of the bio-nanotechnology, many advanced biomedical approaches have been explored for gene delivery. New generations of gene delivery are much effective and safer than viral vectors. [10, 11]

In this chapter, the overview of the gene therapeutic delivery has been discussed, including the recent research progress of physical transfection techniques, viral-based delivery vectors, chemically engineered delivery systems, and bio-inspired vehicles.

## PHYSICAL TRANSFECTION TECHNIQUES

Physical transfection technique includes micro-needle injection, electroporation and hydroporation to make use of the cell permeability to deliver gene therapeutics. The most common physical transfection methods are microinjection and electroporation, while other methods such as hydrodynamics delivery are currently under investigation for clinical application. [9]

Microinjection disrupts the cellular membrane with the microfluidic devices to allow target genes to diffuse into cells. This method is the simplest and safest method, which can apply for specific tissue such as muscle, skin, liver, cardiac muscle, and solid tumors. [12] This method has been considered the highest efficient approach for introducing CRISPR components into cells. Microinjection is not limited by the molecular weight of the cargo and allows for controlled delivery of known quantities of cargos, which improves the control over the off-target effect. However, microinjection is not ideal for in vivo delivery. [9] Yang et al. (2013) used a microscope and a needle to pierce a cell membrane and directly delivered the cargoes within the cell. Cargoes could be either plasmid DNA encoding, both Cas9 protein and the sgRNA or Cas9 protein with sgRNA. [13] Horii et al. (2014) reported that injection of Cas9 mRNA and sgRNA into the zygote cytoplasm and full-term mouse pups with desired modifications. [14] Wu et al. (2013) delivered Cas9 mRNA directly into the cytoplasm by microinjection to correct a cataract-causing mutation in mice. [15] Niu et al. (2014) also disrupted the two genes in cynomolgus monkeys from a single injection of CRISPR mRNA components into one-cell-stage embryos. [16] Long et al. (2014) corrected Duchenne muscular dystrophy (DMD)-causing mutation in mice with the same procedure above. [17]

Electroporation applies an electrical pulse to disrupt the membrane and allows genes to enter cells. This technique involves pulsed high-voltage electrical currents to transiently open the cellular membrane, which would enable components to flow into the cell. This method is less dependent on cell type than other delivery techniques. However, electroporation is generally not suitable for *in vivo* applications. [18] A few groups have reported of CRISPR/Cas9 delivery via electroporation. Hashimoto and Takemoto (2016) used an electroporation chamber to achieve high transfection rate of CRISPR/Cas9 to viable embryos. [19] Specialized electroporation called nucleofection can deliver cargoes directly into the nuclei of mammalian cells. Plasmid encoding both Cas9 and sgRNA had been delivered via nucleofection to correct a cataract-causing mutation in mouse spermatogonial stem cells. [20] Other examples, electroporation was used to edit genes *in vivo*. Zuckermann et al. (2015) injected both Cas9 and sgRNA into the embryo cerebral ventricular zone, followed by electroporation. [21] Recently, Xu et al. used electroporation to deliver Cas9 and gRNA plasmid into skeletal muscle resulting functions of local tissues in a mouse model of Duchenne muscular dystrophy. [22]

Hydrodynamic delivery uses an increase in hydrodynamic pressure to deliver gene editing cargos, typically using the tail vein administration in mice. Increasing the hydrodynamic pressure increases permeability into endothelial and parenchymal cells, allowing naked DNA plasmids and proteins to pass into a cell. [23] Yin et al. (2014) successfully delivered DNA plasmid encoding Cas9 and sgRNA via hydrodynamic delivery to *Fah* mutated mouse hepatocytes *in vivo*. [24] Guan et al. also delivered naked plasmid DNA encoding CRISPR/Cas9 components to a mouse model of hemophilia B using hydrodynamic delivery. [25] Lin et al. (2014) and Zhen et al. (2015) reported delivering of CRISPR/Cas9 component to B virus (HBV) replication and gene expression in HBV-infected mice, respectively. [26, 27] Although with some successes, hydrodynamic delivery is not currently being reported for clinical applications, because they have very low transfection rates, only works on specific types of cells and easy to cause accidental mortality. [23]

## VIRUS-BASED DELIVERY VECTORS

Viral vectors as gene delivery are more efficient than non-viral vectors in various types of cells due to their excellency of overcoming extracellular and intracellular barriers and defense mechanisms of targeted cells. Viral vectors include retroviral, adenoviral, and adenoviral associated vectors, which are all efficient for delivering the genetic materials to the host cells. [28] Viral vectors could be applied to various types of tissues, including liver, eye, skeletal, and cardiac muscle. The broad tropism *in vivo* makes viral-based delivery attractive for the gene delivery. Studies have been applied to *in vitro*, *ex vivo* and *in vivo*. [7] Despite these benefits, viral vectors frequently invoke the host's immune system, which can reduce the effectiveness of gene delivery. [29] Adenovirus causes a very strong immune response of Toll-like receptor (TLR) independent and dependent innate immune system signaling pathways. [30] Moreover, viral glycoproteins can induce the activation of the adaptive immune response. [30]

DNA based viral vectors are generally long-lasting and can integrate into the host genomes. [1] They can deliver genetic materials to the host cells, [31] such as plasmid-based transgene delivery. [32] Although the risk of viral vectors, clinical trials via viral vectors are promising for gene therapies, including cancer, AIDs, neurological disorders such as Alzheimer's disease, Parkinson's, and cardiovascular disorders. [33]

RNA based viral vectors have been developed to deliver RNA to transcribe directly for infectious RNA transcript. RNA based gene delivery is generally considered as transient and not permanent. [34] Currently, RNA-based gene delivery has been made to a patient for HIV with lentiviral vector-modified CD 34 (+) cells. This method has demonstrated the sustained expression of siRNA and ribozyme. [35]

Oncolytic viral vectors are rising treatment for cancer diseases. The goal is to design a virus which can replicate effectively within the host, which focused on cancer research. [36] Oncolytic adenovirus vectors for gene therapy show a promising modality to treat cancer. [37]

## Adenovirus

Adenovirus has been the first DNA virus to enter rigorous therapeutic development due to its genetic stability, high gene transduction efficiency, and well-defined biology. [38] Compared with other viral vectors, adenoviral vectors have significant advantages: the most effective gene delivery system *in vivo*, extensive experience, and versatile developing strategies. [38] In contrast to lentiviral vectors, adenoviruses their DNA does not integrate into the host genome, but transiently express transgene.

However, improvements should be made on immunogenicity, viral longevity, vector packaging capacity, and contamination with a helper virus (HV). [39]

Many groups are using either adenoviral vector for delivery of CRISPR/Cas9 components. For example, the high-capcity adenovirus (HC-AdV)s have been used to effectively deliver CRISPR/Cas9 systems to target tissues or cells. [40] Moreover, HC-AdVs have successfully delivered anti-PD1/PD-L1 immune checkpoint therapeutics for cancer treatment. Recently, Voets et al. (2017) inactivated genes in normal human lung fibroblasts and bronchial epithelial cells by adenovirus. They demonstrated that co-transduction of Adv Cas9 with SMAD3-targeting guide RNAs have genome edited resulting in a reduction of SMAD3 protein expression and nuclear translocation. [41]

## Lentivirus

Both lentivirus and adenovirus can infect dividing and non-dividing cells. However, integrating lentiviral vectors may cause low efficiency and concern about safety, which has a high risk of leukemogenesis. Although making the HIV provirus as integration-deficient as possible, it does not entirely prevent the integration into the host genome. [42] However, only the lentivirus can integrate into the genome, which provides advantageous in the case of CRISPR/Cas9-based editing for limiting off-target effects. [43] Moreover, compared with the AAV, lentivirus has a larger size to pack additionally more space. Lentivirus has been studied for their utility in delivering genes to neural stem cells and within CNS tissues *ex vivo.* Numerous groups have applied lentiviral vectors within animal models for investigating potential therapies for Parkinson disease, Alzheimer disease, and Huntington's disease. [44, 45, 46]

Kabadi et al. (2014) developed a unique lentiviral CRISPR/Cas9 system using Golden Gate synthesis. Their unique Golden Gate design allowed for the expression of one Cas9 and four different sgRNAs on different types of human cells by controlling different promoters. [47]

Recently, IDLV effectively reduced the integrating of the host genome, which may cause low efficiency or safety concern. [48]

## Adeno-Associated Virus

AAV vectors are attractive deliver vector on the context of RNAi delivery as they only need to deliver smaller sized to perform gene silencing. However, on gene editing, low capacity of the gene vector may inhibit the future use of this viral vector. [49] AAV delivery systems are limited by the packaging capacity, which only holds about 5kb in total. Therefore, the AAV delivery system requires the use of at least two vectors to deliver components of gene therapeutics. [50] Long et al. successfully packed the Spcas9 to AAVs and delivered to correct DMD-causing mutations in mice. AAVs were delivered by retro-orbital injection, intraperitoneal or, intramuscular injection and resulted in muscle function enhancement. [17] Dual packaging spCas9 and guide RNA expression cassettes into two separate viral vectors has been reported. [51] The Split cas9 system which two Cas9 halves is packaged into one AAV vector and the Cas9 N-terminal is packaged into a second AAV vector followed by dimerization with a chemically inducible system. [52] Most recent developed AAV CRISPR/Cas9 delivery method was SaCas9 which is a smaller version of Cas9 from *S. aureus* rather than SpCas9 from *Streptococcus pyrogenes*. The SaCas9 is a smaller size of SpCas9 while retaining the same potency. Hence, the smaller size of SaCas9 can be packed within the AAV particles with free space. [53] Ran et al. (2015) have used SaCas9 in AAV vectors to target the cholesterol regulatory gene Pcsk9 and edit the mutations in the DMD gene in adult mice. [54, 55] Shorter variants of Cas9 from *Streptococcus thermophilus* and *Neisseria meningitidis* have also been used for gene editing for packing more efficiently within AAV vector. [56, 57]

Adeno-associated virus delivered shows the promising result to genetic repair of cellular phenotype. Systemic delivery of Cas9 and sgRNAs via AAV vectors have been shown to create exon deletion of the DMD mice resulting in biochemical and functional improvement in skeletal and cardiac muscle. [58, 59] Moreover, AAV delivery of NHEJ has also applied to genetic diseases of the central nervous system such as Huntington's disease, an autosomal dominant form of retinitis pigmentosa, and Leber congenital amaurosis. [10] The strategy to anticipate potential gene therapy is to excise an expanded repeat of a deleterious gene specifically. [60, 61]

There are several approaches to reduce potential genotoxicity associated with persistent expression of nucleases. Merienne N. et al. reported a self-inactivating KamiCas9 system delivered by AAV vector to prevent genotoxicity. [62]

## CHEMICALLY ENGINEERED DELIVERY SYSTEM

Gene therapeutic delivery can be improved by using synthetic nanoparticles composed of chemically engineered materials and conjugates and surface chemistry on the carriers. Compared with viral and lipid-based vector, chemically derived delivery systems are simpler to generate with reproducible sizes and compositions, simpler for modification, and more stable over time.

### Dendrimers

Dendrimers have some advantages than some of the conventional nanocarriers such as liposomes, NPs, and microspheres. The small size of dendrimers, large encapsulation cavities and the capability of swelling make it attractive as nanocarrier. [63] However, cytotoxicity caused by the cationic charge on the surface must be overcome. Currently, dendrimers have been developed to improve its cytotoxicity as biodegradable dendrimers, polyester dendrimers, melamine dendrimers, triazine dendrimers, and poly-L-lysine dendrimers and by surface modification such as PEGylation, carbohydrate coating, acetylation, and amino acid and peptide conjugation. [64] For siRNA delivery, dendrimers are advantageous because researchers can strictly control its structure, size, and the dendrimer's functional groups. [65, 66] Two designed dendrimers called polyamidoamine (PAMAM) and polypropyleneimine (PPI) are currently investigating due its structure amine-terminated, which allows for increased specificity in the release of their contents by the pH-sensitive response. [67, 68]

## Nanogels

Nanogels are attractive delivery system due to combined attributes of both nanoparticles and hydrogels. In addition, nanogels make the carriers biocompatibility, high stability, and high loading capacity. Hence, siRNA-loaded nanogels can be prepared by polymerization and chemical crosslinking. [67] For example, Li et al. (2016) investigated the potential use of ethylenediamine (ED)-functionalized low molecular-weight PGMA nanogels (PGED-NGs) as effective siRNA and pDNA carrier. These crosslinked nanogels had successfully transfected pDNA and target-specific intracellular delivery of MALAT1 siRNA into hepatoma cells. [69] By designing the nanogel with nano-scaled dimension and morphology, nanogels can achieve the stimuli-controlled release of bioactive compounds. For example, F. Reisbeck et al. (2017) developed pH-sensitive nanogels for siRNA mediated gene silencing. Novel photochemical polymerization along with siRNA nanogels was developed to prolonged gene silencing. [70] For gene editing using nanogel, Hong et al. (2017) delivered CRISPR/Cas9 components via liposome-templated hydrogel nanoparticles to a tumor cell. [71]

## Gold Nanoparticles

Gold nanoparticles have shown to be attractive material due to its inert nature, ease of preparation, and unlimited surface characterization properties. Moreover, golds are colloidally stable, minimizing the risk of early clearance than liposomes. These characteristics make them readily used in *in vitro*, *ex vivo*, and *in vivo* settings for gene therapy. [72] *In vivo*, gold nanoparticles have been reported as low toxicity and minimal off-target genomic toxic at low concentration. [73] However, Dykman and Khlebtsov reviewed that high concentrations of gold nanoparticles were shown to stimulate immune cytokine production. In general, the toxicity of gold nanoparticles depends on their size, shape and surface chemistry. Therefore, each functionalized gold nanoparticles should investigate their immune response. [74] The versatility of surface chemistry on gold nanoparticles has made it an attractive carrier for siRNA delivery. For instance, gold nanoparticles can be modified to have pH-sensitivity, which could release their contents into a target region at acidic tumor environment. [75] Gu et al. reported a modified PEG with cell-penetrating peptide (CPP) on a gold core. Moreover, cationic polymers (chitosan and poly-L-lysis) and cysteine-coated AuNPs also showed excellent stability in blood circulation and effective delivery of gene therapeutics. [76] Son et al. developed a gold-based nanoconstruct to deliver RNAi materials with various geometries. This nanoconstruct showed precise conjugation and separation of a designated number of therapeutic siRNAs onto AuNP. [77] Mout et al. (2017) used gold nanoparticles to co-assemble with engineered Cas9 protein and sgRNA into nano-assemblies. This construct provided a direct platform for multiple in vitro applications. [78] Lee et al. (2017) reported CRISPR-gold can deliver Cas9 protein, gRNA, and donor DNA both *in vitro* and *in vivo*. Gold nanoparticle-mediated Cas9 RNP delivery was successfully delivered to edit gene and HDR *in vivo*. Systemic injection of gold nanoparticle-conjugated Cas9 RNP and donor DNA into a DMD mouse model has been shown to correct the dystrophin reading frame. [79] Lee et al. (2018) also demonstrated that the intracranial injection of CRISPR-Gold can edit genes in the brains of adult mice. It was shown that the CRISPR-Gold can target the metabotropic glutamate receptor 5 (mGluR5) gene and reduce the expression level of mGlur5 in the striatum after intracranial injection. [80]

## Mesoporous Nanomaterials

Mesoporous nanomaterials (MSNs) for gene delivery also show promising results. MSNs are highly negative charged materials, which required modifications to express net positive charge by methods including amination-modification, metal cations co-delivered vector, and cationic polymer functionalization to bind to negatively charged nucleic acids. Positive charged MSNs have ordered porous structure and large pore volume and surface area with excellent biocompatibility. [81] MSNs for siRNA delivery would be mesoporous silica-based nanoparticles. They can encapsulate more molecules compared to conventional carriers. Moreover, MSNs are stable due to their iron oxide framework and have been reported to escape the endosome. However, more research is required on safety, biodegradability, pharmacokinetics, and biodistribution of MSNs. [82] Recently, chemically engineered MSNs; pH-sensitive linkers and polyelectrolyte gatekeepers, supramolecular-nanovalves and acid-decomposable inorganic gatekeepers have been developed to improve MSN delivery for siRNA delivery. [83, 84, 85, 86] For CRISPR Cas9 delivery, using MSNs has not been reported. There are many barriers to overcome, mainly the packaging of large cargoes. CRISPR Cas9 components as RNP complex or mRNA or as DNA plasmids are larger than reported to load within the particles. In addition, more *in vivo* research are required to systematically investigated on biocompatibility.

## Polysaccharide Polythyleminime

Polysaccharides polythyleminime (PEI) is a cationic polymer which has high-density amine groups. Delivery of PEI exerts 'protein sponge effect' that eventually stops the acidification of endosomal pH. The proton sponge effect makes the influx of chloride into the polymer and increases the osmotic pressure, which swells and ruptures the endosomal membrane. [87] Due to this facilitating efficient plasmid DNA packing and pH-buffering ability, PEI or poly(L-lysine), PLL has been the common polymeric vehicles for DNA delivery. However, branched PEI or unmodified PEI is cytotoxic and showing low transfection rate due to its high positive charge. [88] Therefore, an appropriate balance of linear and branched structure is necessary to be less toxic and more effective transfection. [89] PEI has also been a popular material for siRNA delivery. Goyal et al. reported linear PEI nanoparticles have efficiently delivered plasmid DNA or siRNA in vitro and in vivo. [90] Gene transfer via PEI has been reported by Zuckermann et al. PEI/Cas9/sgRNA plasmid vehicle was used to target tumor suppressor genes in the mouse brain. [91] Zhen et al. also used PEI to deliver CRISPR plasmid DNA to inhibit HBV replication and gene expression in HBV-infected mice. [21]

## Poly-L-Lysine (PLL)

Currently, Poly L-lysine (PLL) has been widely used as a non-viral gene vector. PLL is a synthetic polypeptide with amino-acid lysine as a repeat unit and is biodegradable, which can electrostatically interact with a negative biomolecule such as siRNA. [27] However, PLL shows a low level of transfection efficiency and suffers from immunogenicity caused by its amino-acid backbone. [92] Therefore, to overcome toxicity, PEG-coated PLL has been developed to prolong its lifetime in blood circulation and tumoral accumulation. [93] Harashima group developed multifunctional-envelope-type nanodevice called MENDs which composed of condensed plasmid DNA, a PLL core, and a lipid film shell. [94] Recently, a tetra-lamellar MEND (T-MEND) was developed and successfully deliver pDNA and siRNA. [95] Therefore, PLL could become promising polypeptide to complex with Cas9 plasmid DNA, forming a multifunctional envelope-type nanodevice (MEND). [96]

## Semiconductor Quantum Dots

Semiconductor Quantum Dots have become attractive to researchers as well, because of the unique photoelectric properties of quantum dots and good biocompatibility. Nanosized quantum dots have unique characteristics such as high specific surface area, tunable fluorescence wavelength, easy surface modification and low toxicity, which makes attractive as vectors for target transport and biological imaging. [9] siRNA delivery via quantum dots has been reported into cells and small animals. For example, Cadmium sulphoselenide/Zinc sulfide quantum dots (CdSSe/ZnS QDs)-based nanocarriers were developed for *in vitro* gene delivery. These quantum dots functionalized with PEI to form stable nanoplex and loaded with siRNA, which specifically targets human telomerase reverse transcriptase(TERT). This result in suppression of proliferation of glioblastoma cells. [97] However, quantum dots as potential therapeutics for human drug delivery vehicles are unclear. For example, some *in vitro* studies show that QDs are harmful to the cells when high doses of QDs are exposed. [98]


## Superparamagnetic Iron Oxide Nanoparticles (SPIONs)

Superparamagnetic Iron Oxide Nanoparticles called SPIONs are also rising inorganic particles due to its characteristics of biodegradability, biocompatibility, and nanotoxicity. These particles exhibit the phenomenon of "superparamagnetism" which on an external magnetic field, they become magnetized up to their saturation magnetization. [99] Currently, there is a research demonstration on how external magnetic field could be applied to improve the transfection rate. SPIONs coated with cationic polymers were demonstrated to increase the magnetofection efficiency of therapeutic genes. By applying an external magnetic field, PEI-coated SPION can be reduced to be free diffusion hence enhanced transfection efficiency. Namgung et al. designed thermally crosslinked superparamagnetic nanoparticles containing polyethylene glycol, branched PEI and DNA of interest. This hybrid nanoparticles have successfully inhibited the expression of plasminogen activator inhibitor-1 by high transfection efficiency in vascular endothelial cell. [100] Hwang et al. developed a magnetically guided adeno-associated virus delivery system for enhancing gene delivery to cell line. [101] However, toxicity is a significant concern over SPIONs once it is degraded in the body.

## Carbon Nanotubes

Carbon nanotubes (CNTs) made of one or more sheets of graphene shaped in a cylindrical structure which have small size and chemical inertness. This creates an attractive material for the delivery of DNA. [102] For instance, carbon nanotubes have recently shown a promising result for cancer treatment, as a carrier for siRNA delivery. [103] Carbon nanotubes categorized as single-walled and multiwalled, which the single walled CNTs functionalized with *-CONH-(CH$_2$)6-NH$_3$)* + *Cl$^-$* act as siRNA carriers. The siRNA is released from the nanotube side-wall to function as gene silencing. [104, 105] Wang et al. succeeded in delivering siRNA using PEI and tumor targeting peptide coated carbon nanotubes to increase transfection efficiency. As a result, carbon nanotube delivered siRNA against hTERT by modifying the surface with tumor-targeting peptide and PEI. [106] Recently, specific carbon nanotubes with nanoneedles are being studied for the delivery system due to their cell-death inducing activity. This CNT with their nanoneedle structure could independently translocate into cytoplasm without causing necrobiosis. [107, 108] Also manipulated carbon nanomaterial called nano-diamond were also used for the delivery siRNA into cells efficiently and downregulate the expression of desired proteins. [109] Furthermore, developed form of nano-diamond *NHCONH(CH$_2$)$_2$NH-VDGR*, containing nano-diamond successfully delivered surviving-siRNA and downregulated the expression of the surviving protein resulting in tumor apoptosis. [110]

## BIO-INSPIRED VEHICLES

Extracellular vesicles and protein-based vehicles have been noted as biomimetically inspired approaches, which can mediate intercellular RNA transfer physiologically.

## Exosome-Based Vehicles

Exosomes are naturally produced lipid vesicles, which could be used as siRNA carriers. There are several advantages because exosomes present nanoscopic sizes, low immunogenicity, high biocompatibility, encapsulation of various cargoes, and the ability to overcome the biological barrier. [111] Yang et al. (2016) investigated whether brain endothelial cell-derived exosomes could deliver siRNA across the blood-brain barrier in zebrafish and confirm it could be potentially used as a natural carrier for the brain delivery of exogenous siRNA. However, it is limited by their low efficiency in encapsulating of large nucleic acids. [112] Hence, Lin et al. (2018) developed a kind of hybrid exosomes with liposomes via simple incubation. The resultant hybrid nanoparticles efficiently encapsulated large nucleic acids such as plasmids and CRISPR-Cas9 expression vectors. This hybrid nanoparticle can successfully deliver the CRISPR-Cas9 system in mesenchymal stem cells (MSCs). [113]

## Protein-Based Vehicles

The protein-based vehicles offer significant therapeutic potential due to their addressability and tolerates manipulation. However, proteins are inherently unstable and complex molecules. *In vivo*, they are susceptible to proteases in circulation. They are also highly bioactive, which produce unintended effects. [114] In the case of protein-based RNAi, delivery barriers include instability in circulation, rapid clearance, inability to circumvent endosomal degradation, and nonspecific delivery. [115] Recently, Yang et al. enhanced the delivery of p19 protein for dsRNA through yeast display by enhancing the binding affinity. [116]

## Polysaccharide-Based Vehicles

Chitosan is composed of glucosamine/acetylglucosamine, which is a cationic, linear polysaccharide. This has become an attractive vector for siRNA due to its biocompatibility, low cytotoxicity, and generally lack of immunogenicity. [117] High positively charged chitosan was able to interact with negatively charged siRNAs and protect it. However, the variability in composition and molecular weight on chitosan have been a significant challenge to chitosan. Moreover, chitosan transfection efficiency is sensitive to its surroundings, such as pH. [118] Currently, chemical modification on the amino group and the primary hydroxyl group of chitosan improved the siRNA delivery. Other modifications, such as PEGylation has optimized chitosan as a delivery vector. [119]

## Peptide-Based Vector

Peptides are widely used as non-viral gene vectors due to their low cytotoxicity and immunogenicity profiles. [120] CPPs are a series of short amino acids that are polycationic, amphiphilic, or non-polar in nature. [121] Conjugating CPPs with each cargo can facilitate uptake into the cell, which works well for *in vitro* and *ex vivo,* but CPPs are less successful to be utilized to deliver components *in vivo.* CPPs can carry siRNA duplexes through electrostatic interactions by either covalent or noncovalent linkage. However, when positively-charged CPPs are covalently linked to negatively charged siRNAs, the delivery efficacy decreases. Additionally, CPPs cause cytotoxicity or inducing immunogenicity *in vivo.* [122] Ramakrishna et al. reported that CPP-Cas9 and CPP-sgRNA, as CPP conjugation can enhance protein delivery, resulted in gene editing of multiple human cell lines. [123] Axford et al. (2017) demonstrated that CPP-delivered CRISPR/Cas9 RNP cellular and sub-cellular localizations using a confocal microscope. [124] However, CPPs show low efficiency of the gene editing due to absent of protection from the protease degradation.

Peptide itself can be used as vectors for the delivery of oligonucleotides, plasmid DNA, and siRNAs. Peptides have been reported as very useful building blocks for designing nanostructures because of their structure folding. If the *in vivo* circulation time can be extended, peptides may be promising gene delivery vehicles. Recently, researchers hope to integrate functional peptide sequences with non-natural amino acids or functional moieties to design peptide-based nanostructures. However, the major barriers reported are short circulation half-lives, poor chemical, and physical stability in serum and low DNA binding ability. Hence, self-assembled peptide-based nanostructures are currently investigated to improve transfection efficiency. [125] Recently, Stupp and coworkers developed a peptide-based vehicle composed of a self-assembling domain with a specific RNA recognition that could combine siRNA resulting to generate sub-micrometer supramolecular particles. These siRNA/peptide nanoparticles can be delivered to neural cells with high transfection efficiency. [123]

## DNA-Based Nanostructure

The DNA-nanostructures are intrinsic biocompatibility and biodegradability, which makes highly intriguing delivery vehicles. [126] The unique property of DNA nanostructure-based vehicle is the modularity, meaning the size and position of modification can be precisely controlled at the nanoscale. Thus, the shape and flexibility of the DNA object can be fine-tuned. In general, DNA bio-inspired materials are considered safe on the immunogenicity. Therefore, DNA nanostructures can be easily programed to carry siRNAs with DNA-RNA hybridization. Lee et al. designed a self-assembled DNA tetrahedral nanostructure for siRNA delivery *in vivo* to silence target genes in tumors. [127] In 2016, Sleiman et al. made DNA "nanosuitcase" that could encapsulate a siRNA to protect siRNA and release cargoes on demand. [128] Mirkin and co-workers constructed nanoparticle-assisted DNA nanostructures for siRNA delivery. This platform composed of gold nanoparticles with assembling spherical nucleic acid (SNA). SiRNA-loaded SNA can penetrate the blood-brain barrier and blood-tumor barrier to disseminated through xenogenic glioma explants. [129] Microsponges for siRNA delivery was developed by Hammond and co-workers. The microsponge can self-assemble into nanoscale-pleated sheets of hairpin RNAi by using rolling-circle transcription. [130] Furthermore, multi-RNAi microsponge platform has been improved to the simultaneous controlled delivery of multiple siRNAs. They demonstrated that the microsponge-based siRNA delivery could potentially be used for the treatment of cancer, genetic disorder, and viral infection. [131] In 2017, Leong et al. designed and assembled a DNA star motif which can carry three miRNA molecules. [132] Nahar et al. also assembled a programmable anti-miR branched DNA nanostructure carrying single-stranded anti-miRNA overhangs. These studies indicated that DNA nanostructure can be used for miRNA-based cancer therapies. [133]

DNA nanostructures also can be applied for delivery of gene editing components. Gu et al. (2015) developed a DNA-based self-assembled nanoparticle called "DNA nanoclew" coated with cationic polymer polyethylenimine. The nanoclew is self-assembled by rolling circle amplification (RCA) with palindromic sequences. This nanoclews efficiently delivered Cas9/sgRNA ribonucleoprotein complex. DNA nanoclew delivered the Cas9 protein and show transient expression of gene editing, but still, involve no viral components. However, it did not have a cell-specific targeting ligand and did not include environmentally responsive elements to release of CRISPR-Cas9. Moreover, further test is required particularly on the potential immunogenicity of DNA nanoclews. [134, 135]


**Lipid Based Vehicle**

Liposomes are often used as a delivery vehicle for RNAi, including siRNAs. Recently, the first RNAi therapeutics via liposome (Onpattro) has been approved by the FDA (August 2018). This approach generally involves cationic lipids to overcome the negative charge of the siRNAs. Hence, lipoplexes have been developed, combining an appropriate ratio of cationic liposomes to siRNAs. [136] However, previously the potential clinical use of cationic liposomes has its limitation due to their instability, rapid system clearance, toxicity, and induction of immunostimulatory response. [137] Therefore, many optimizations have been designed to overcome these limitations. In addition, various lipid derivatives functionalized liposomal to target for the delivery of DNA actively. [138] Generally, many different targeting ligands such as sugars, peptides, proteins, aptamers, or antibodies have been investigated. [139]

Liposome that contains polymers called lipopolyplexes can deliver siRNA effectively due to polymeric components have a natural affinity for RNA. For example, stable nucleic acid-lipid particles (SNAPs) have been developed to deliver siRNAs efficiently. Morrissey et al. made stable nucleic acid-lipid particles to transport siRNA against HBV, which prohibit the virus in mice. [140] Research on lipopolyplexes is currently being conducted to investigate their usefulness as an alternative to viral vectors for the treatment of Parkinson's disease, because it can cross the blood-brain barrier and target diseased brain cells specifically. [141]

For gene editing delivery, engineering of negatively charged TALE protein or RNPs has been loaded to cationic lipids. These show higher gene editing efficiency at micromolar concentration. To increase the probability of nuclear gene transfer, substances, including nuclear localization sequences (NLS), were investigated. [142]

Zuris et al. reported that Cas9 protein and sgRNA targeting EGFP encapsulated in cationic liposomes have been delivered to hair cells in the inner ear of a GFP reporter showing 13% loss. [143] Recently, Zhang et al. (2017) developed a lipid-based delivery vehicle called polyethylene glycol phospholipid-modified cationic lipid nanoparticle (PLNP) which deliver a CRISPR/Cas9 system and treated melanoma tumor-bearing mice. [144]

## CURRENT AND FUTURE DEVELOPMENT

Multiple gene delivery systems and extensive ongoing studies to accomplish targeted delivery of gene silencing and editing has been reviewed in this chapter. Gene silencing, especially the siRNA therapeutics delivery, has been pursued actively due to their high specificity and unlimited therapeutic targets. However, naked siRNA, as inherited, show instability in the blood. Hence, challenges in designing delivery systems to target sites have limited its potential. Continued research to overcome these barriers would expand the potential use of RNA drugs for genetic infectious and chronic disease and will accelerate the trend toward precision medicine.

The full potential of CRISPR/Cas9 should overcome many challenges. The off-target cutting of CRISPR/Cas9 remains a problem. Advanced gene therapeutic delivery vehicles presenting high-specificity may limit off-target effects. Every delivery method has both advantages and disadvantages. Development of new delivery approaches considering both safety and specificity will make a meaningful impact on the field. With improved design and delivery, new genetic treatment will cure genetic diseases that are currently untreatable.

## CONCLUSION

New generation on various nucleic acid delivery platforms improves research and applications of gene therapy. While there are major obstacles to be solved, such as off-target and immunogenicity, many promising results for each delivery platform have been reported in this field. Safer and more effective gene therapeutic delivery systems have been developed, which positively impacts the gene therapy field. Further development of nucleic acid delivery platforms will ultimately realize successful gene therapies.

## REFERENCES

[1]   Gonçalves, G. A. R. & Paiva, R. de M. A. Gene therapy: advances, challenges and perspectives. *Einstein (São Paulo)* (2017). doi:10.1590/s1679-45082017rb4024.

[2]    Tatiparti, K., Sau, S., Kashaw, S. & Iyer, A. siRNA Delivery Strategies: A Comprehensive Review of Recent Developments. *Nanomaterials* 7, 77 (2017).

[3]    Ho, W., Zhang, X. Q. & Xu, X. Biomaterials in siRNA Delivery: A Comprehensive Review. *Adv. Healthc. Mater.* 5, 2715–2731 (2016).

[4]    Marquez, A. R., Madu, C. O. & Lu, Y. An Overview of Various Carriers for siRNA Delivery. *Oncomedicine* 3, 48–58 (2018).

[5]    Qureshi, A., Tantray, V. G., Kirmani, A. R. & Ahangar, A. G. A review on current status of antiviral siRNA. *Rev. Med. Virol.* 28, 1–11 (2018).

[6]    Wittrup, A. & Lieberman, J. Knocking down disease: A progress report on siRNA therapeutics. *Nat. Rev. Genet.* 16, 543–552 (2015).

[7]    Kotterman, M. A., Chalberg, T. W. & Schaffer, D. V. Viral Vectors for Gene Therapy: Translational and Clinical Outlook. *Annu. Rev. Biomed. Eng.* (2015). doi:10.1146/annurev-bioeng-071813-104938.

[8]    Pecot, C. V., Calin, G. A., Coleman, R. L., Lopez-Berestein, G. & Sood, A. K. RNA interference in the clinic: Challenges and future directions. *Nat. Rev. Cancer* 11, 59–67 (2011).

[9]    Lino, C. A., Harper, J. C., Carney, J. P. & Timlin, J. A. Delivering crispr: A review of the challenges and approaches. *Drug Deliv.* 25, 1234–1257 (2018).

[10]   Liu, C., Zhang, L., Liu, H. & Cheng, K. Delivery strategies of the CRISPR-Cas9 gene-editing system for therapeutic applications. *J. Control. Release* 266, 17–26 (2017).

[11]   Yin, H., Kauffman, K. J. & Anderson, D. G. Delivery technologies for genome editing. *Nat. Rev. Drug Discov.* 16, 387–399 (2017).

[12]   Mellott, A. J., Forrest, M. L. & Detamore, M. S. Physical non-viral gene delivery methods for tissue engineering. *Annals of Biomedical Engineering* (2013). doi:10.1007/s10439-012-0678-1.

[13]   Wang, H. *et al.* One-step generation of mice carrying mutations in multiple genes by CRISPR/cas-mediated genome engineering. *Cell* (2013). doi:10.1016/j.cell.2013.04.025.

[14] Horii, T. et al. Validation of microinjection methods for generating knockout mice by CRISPR/Cas-mediated genome engineering. *Sci. Rep.* (2014). doi:10.1038/srep04513.

[15] Wu, Y. et al. Correction of a genetic disease in mouse via use of CRISPR-Cas9. *Cell Stem Cell* (2013). doi:10.1016/j.stem.2013. 10.016.

[16] Niu, Y. et al. Generation of gene-modified cynomolgus monkey via Cas9/RNA-mediated gene targeting in one-cell embryos. *Cell* (2014). doi:10.1016/j.cell.2014.01.027.

[17] Long, C. et al. Prevention of muscular dystrophy in mice by CRISPR/Cas9-mediated editing of germline DNA. *Science (80-. ).* (2014). doi:10.1126/science.1254445.

[18] Young, J. L. & Dean, D. A. Electroporation-Mediated Gene Delivery. *Adv. Genet.* (2015). doi:10.1016/bs.adgen.2014.10.003.

[19] Hashimoto, M., Yamashita, Y. & Takemoto, T. Electroporation of Cas9 protein/sgRNA into early pronuclear zygotes generates non-mosaic mutants in the mouse. *Dev. Biol.* (2016). doi:10.1016/j.ydbio. 2016.07.017.

[20] Wu, Y. et al. Correction of a genetic disease by CRISPR-Cas9-mediated gene editing in mouse spermatogonial stem cells. *Cell Res.* (2015). doi:10.1038/cr.2014.160.

[21] Zuckermann, M. et al. Somatic CRISPR/Cas9-mediated tumour suppressor disruption enables versatile brain tumour modelling. *Nat. Commun.* 6, 1–9 (2015).

[22] Xu, L. et al. CRISPR-mediated genome editing restores dystrophin expression and function in mdx mice. *Mol. Ther.* 24, 564–569 (2016).

[23] Bonamassa, B., Hai, L. & Liu, D. Hydrodynamic gene delivery and its applications in pharmaceutical research. *Pharmaceutical Research* (2011). doi:10.1007/s11095-010-0338-9.

[24] Yin, H. et al. Genome editing with Cas9 in adult mice corrects a disease mutation and phenotype. *Nat. Biotechnol.* (2014). doi:10.1038/nbt.2884.

[25] Guan, Y. et al. CRISPR/Cas9-mediated somatic correction of a novel coagulator factor IX gene mutation ameliorates hemophilia in mouse. *EMBO Mol. Med.* (2016). doi:10.15252/emmm.201506039.

[26] Lin, S. R. et al. The CRISPR/Cas9 system facilitates clearance of the intrahepatic HBV templates in vivo. *Mol. Ther. - Nucleic Acids* (2014). doi:10.1038/mtna.2014.38.

[27] Zhen, S. et al. Harnessing the clustered regularly interspaced short palindromic repeat (CRISPR)/CRISPR-associated Cas9 system to disrupt the hepatitis B virus. *Gene Ther.* 22, 404–412 (2015).

[28] Walther, W. & Stein, U. Viral vectors for gene transfer: a review of their use in the treatment of human diseases. *Drugs* (2000). doi:10.2165/00003495-200060020-00002.

[29] Ibraheem, D., Elaissari, A. & Fessi, H. Gene therapy and DNA delivery systems. *International Journal of Pharmaceutics* (2014). doi:10.1016/j.ijpharm.2013.11.041.

[30] Nayak, S. & Herzog, R. W. Progress and prospects: Immune responses to viral vectors. *Gene Therapy* (2010). doi:10.1038/gt.2009.148.

[31] Wivel, N. A. & Wilson, J. M. Methods of gene delivery. *Hematol. Oncol. Clin. North Am.* 12, 483–501 (1998).

[32] Crooke, S. T. An overview of progress in antisense therapeutics. *Antisense Nucleic Acid Drug Dev.* (1998). doi:10.1089/oli.1. 1998.8.115.

[33] Patil, S. D., Rhodes, D. G. & Burgess, D. J. DNA-based therapeutics and DNA delivery systems: A comprehensive review. *AAPS J.* (2005). doi:10.1208/aapsj070109.

[34] Mogler, M. A. & Kamrud, K. I. RNA-based viral vectors. *Expert Review of Vaccines* (2014). doi:10.1586/14760584.2015.979798.

[35] Digiusto, D. L. et al. RNAms Gene Therapy.pdf. 2, (2011).

[36] Howells, A., Marelli, G., Lemoine, N. R. & Wang, Y. Oncolytic Viruses—Interaction of Virus and Tumor Cells in the Battle to Eliminate Cancer. *Front. Oncol.* (2017). doi:10.3389/ fonc.2017.00195.

[37] Choi, J. W., Lee, Y. S., Yun, C. O. & Kim, S. W. Polymeric oncolytic adenovirus for cancer gene therapy. *J. Control. Release* (2015). doi:10.1016/j.jconrel.2015.10.009.

[38] Crystal, R. G. Adenovirus: The First Effective In Vivo Gene Delivery Vector. *Hum. Gene Ther.* (2014). doi:10.1089/hum. 2013.2527.

[39] Lee, C. S. et al. Adenovirus-mediated gene delivery: Potential applications for gene and cell-based therapies in the new era of personalized medicine. *Genes Dis.* 4, 43–63 (2017).

[40] Schiwon, M. et al. One-Vector System for Multiplexed CRISPR/Cas9 against Hepatitis B Virus cccDNA Utilizing High-Capacity Adenoviral Vectors. *Mol. Ther. - Nucleic Acids* 12, 242–253 (2018).

[41] Voets, O. et al. Highly efficient gene inactivation by adenoviral CRISPR/Cas9 in human primary cells. *PLoS One* 12, 1–20 (2017).

[42] Escors, D. & Breckpot, K. Lentiviral vectors in gene therapy: Their current status and future potential. *Archivum Immunologiae et Therapiae Experimentalis* (2010). doi:10.1007/s00005-010-0063-4.

[43] Philpott, N. J. et al. Efficient Integration of Recombinant Adeno-Associated Virus DNA Vectors Requires a p5-rep Sequence in cis. *J. Virol.* (2002). doi:10.1128/jvi.76.11.5411-5421.2002.

[44] Mandel, R. J., Burger, C. & Snyder, R. O. Viral vectors for in vivo gene transfer in Parkinson's disease: Properties and clinical grade production. *Exp. Neurol.* (2008). doi:10.1016/j.expneurol.2007. 08.008.

[45] Peng, K. A. & Masliah, E. Lentivirus-expressed siRNA vectors against Alzheimer disease. *Methods Mol. Biol.* (2010). doi:10.1007/978-1-60761-533-0_15.

[46] Cambon, K. et al. Preclinical Evaluation of a Lentiviral Vector for Huntingtin Silencing. *Mol. Ther. - Methods Clin. Dev.* (2017). doi:10.1016/j.omtm.2017.05.001.

[47] Kabadi, A. M., Ousterout, D. G., Hilton, I. B. & Gersbach, C. A. Multiplex CRISPR/Cas9-based genome engineering from a single lentiviral vector. *Nucleic Acids Res.* 42, 1–11 (2014).

[48] Wang, Y., Wang, Y., Chang, T., Huang, H. & Yee, J. K. Integration-defective lentiviral vector mediates efficient gene editing through homology-directed repair in human embryonic stem cells. *Nucleic Acids Res.* 45, 1–12 (2017).

[49] Samulski, R. J. & Muzyczka, N. AAV-Mediated Gene Therapy for Research and Therapeutic Purposes. *Annu. Rev. Virol.* (2014). doi:10.1146/annurev-virology-031413-085355.

[50] Wu, Z., Yang, H. & Colosi, P. Effect of genome size on AAV vector packaging. *Mol. Ther.* (2010). doi:10.1038/mt.2009.255.

[51] Swiech, L. et al. In vivo interrogation of gene function in the mammalian brain using CRISPR-Cas9. *Nat. Biotechnol.* 33, 99–103 (2014).

[52] Centre, S. C., Life, V. & Initiative, S. C. *A split-Cas9 architecture for inducible genome editing and transcription modulation.* 33, 139–142 (2015).

[53] Marraffini, L. A. The CRISPR-Cas system of Streptococcus pyogenes: function and applications. Streptococcus pyogenes: Basic Biology to Clinical Manifestations. *[Internet]. U.S. National Library of Medicine, 7 Apr.2016. www.ncbi.nlm.nih.gov/books/NBK355562/.* (2016).

[54] Ran, F. A. et al. In vivo genome editing using Staphylococcus aureus Cas9. *Nature* (2015). doi:10.1038/nature14299.

[55] Nelson, C. E. et al. In vivo genome editing improves muscle function in a mouse model of Duchenne muscular dystrophy. *Science (80-. ).* (2016). doi:10.1126/science.aad5143.

[56] Müller, M. et al. Streptococcus thermophilus CRISPR-Cas9 systems enable specific editing of the human genome. *Mol. Ther.* (2016). doi:10.1038/mt.2015.218.

[57] Lee, C. M., Cradick, T. J. & Bao, G. The neisseria meningitidis CRISPR-Cas9 system enables specific genome editing in mammalian cells. *Mol. Ther.* (2016). doi:10.1038/mt.2016.8.

[58] Bengtsson, N. E. et al. Corrigendum: Muscle-specific CRISPR/Cas9 dystrophin gene editing ameliorates pathophysiology in a mouse

model for Duchenne muscular dystrophy. *Nat. Commun.* 8, 16007 (2017).

[59]  Min, Y. L. et al. CRISPR-Cas9 corrects Duchenne muscular dystrophy exon 44 deletion mutations in mice and human cells. *Sci. Adv.* (2019). doi:10.1126/sciadv.aav4324.

[60]  Shin, J. W. et al. Permanent inactivation of Huntington's disease mutation by personalized allele-specific CRISPR/Cas9. *Hum. Mol. Genet.* (2016). doi:10.1093/hmg/ddw286.

[61]  Ruan, G. X. et al. CRISPR/Cas9-Mediated Genome Editing as a Therapeutic Approach for Leber Congenital Amaurosis 10. *Mol. Ther.* 25, 331–341 (2017).

[62]  Merienne, N. et al. The Self-Inactivating KamiCas9 System for the Editing of CNS Disease Genes. *Cell Rep.* 20, 2980–2991 (2017).

[63]  Tomalia, D. A. et al. A New Class of Polymers: Starburst-Dendritic Macromolecules. *Polym. J.* (1985). doi:10.1295/polymj.17.117.

[64]  Kesharwani, P. et al. Dendrimer nanohybrid carrier systems: an expanding horizon for targeted drug and gene delivery. *Drug Discov. Today* 23, 300–314 (2018).

[65]  Abbasi, E. et al. Dendrimers: Synthesis, applications, and properties. *Nanoscale Research Letters* (2014). doi:10.1186/1556-276X-9-247.

[66]  Mintzer, M. A. & Grinstaff, M. W. Biomedical applications of dendrimers: A tutorial. *Chemical Society Reviews* (2011). doi:10.1039/b901839p.

[67]  Tang, Y. et al. Efficient in vitro siRNA delivery and intramuscular gene silencing using PEG-modified PAMAM dendrimers. *Mol. Pharm.* (2012). doi:10.1021/mp3001364.

[68]  Taratula, O., Savla, R., He, H. & Minko, T. Poly(propyleneimine) dendrimers as potential siRNA delivery nanocarrier: from structure to function. *Int. J. Nanotechnol.* (2010). doi:10.1504/ijnt.2011.037169.

[69]  Soni, G. & Yadav, K. S. Nanogels as potential nanomedicine carrier for treatment of cancer: A mini review of the state of the art. *Saudi Pharmaceutical Journal* (2016). doi:10.1016/j.jsps.2014.04.001.

[70] Li, R. Q. et al. Well-defined reducible cationic nanogels based on functionalized low-molecular-weight PGMA for effective pDNA and siRNA delivery. *Acta Biomater.* (2016). doi:10.1016/j.actbio. 2016.06.006.

[71] Dimde, M. et al. Defined pH-sensitive nanogels as gene delivery platform for siRNA mediated: In vitro gene silencing. *Biomater. Sci.* 5, 2328–2336 (2017).

[72] Chen, Z. et al. Targeted Delivery of CRISPR/Cas9-Mediated Cancer Gene Therapy via Liposome-Templated Hydrogel Nanoparticles. *Adv. Funct. Mater.* 27, 1–9 (2017).

[73] Sau, T. K. & Goia, D. V. Biomedical applications of gold nanoparticles. in *Fine Particles in Medicine and Pharmacy* (2011). doi:10.1007/978-1-4614-0379-1_4.

[74] Zhang, X. D. et al. Toxicologic effects of gold nanoparticles in vivo by different administration routes. *Int. J. Nanomedicine* (2010). doi:10.2147/IJN.S8428.

[75] Dykman, L. & Khlebtsov, N. Gold nanoparticles in biomedical applications: Recent advances and perspectives. *Chem. Soc. Rev.* 41, 2256–2282 (2012).

[76] Li, S. et al. pH-responsive targeted gold nanoparticles for in vivo photoacoustic imaging of tumor microenvironments . *Nanoscale Adv.* 1, 554–564 (2018).

[77] Li, S. et al. Mechanism of Cellular Uptake to Optimized AuNP Beacon for Tracing mRNA Changes in Living Cells. *Part. Part. Syst. Charact.* (2018). doi:10.1002/ppsc.201700331.

[78] Son, S. et al. Antitumor therapeutic application of self-assembled RNAi-AuNP nanoconstructs: Combination of VEGF-RNAi and photothermal ablation. *Theranostics* (2017). doi:10.7150/thn o.16042.

[79] Mout, R. & Rotello, V. Cytosolic and Nuclear Delivery of CRISPR/Cas9-ribonucleoprotein for Gene Editing Using Arginine Functionalized Gold Nanoparticles. *Bio-Protocol* 7, 1–6 (2017).

[80] Lee, K. et al. Nanoparticle delivery of Cas9 ribonucleoprotein and donor DNA in vivo induces homology-directed DNA repair. *Nat. Biomed. Eng.* 1, 889–901 (2017).

[81] Lee, B. et al. Nanoparticle delivery of CRISPR into the brain rescues a mouse model of fragile X syndrome from exaggerated repetitive behaviours. *Nat. Biomed. Eng.* 2, 497–507 (2018).

[82] Luo, G. F. et al. Multifunctional enveloped mesoporous silica nanoparticles for subcellular co-delivery of drug and therapeutic peptide. *Sci. Rep.* (2014). doi:10.1038/srep06064.

[83] Pinese, C. et al. Sustained delivery of siRNA/mesoporous silica nanoparticle complexes from nanofiber scaffolds for long-term gene silencing. *Acta Biomater.* 76, 164–177 (2018).

[84] Chen, M. et al. A pH-responsive polymer/mesoporous silica nano-container linked through an acid cleavable linker for intracellular controlled release and tumor therapy in vivo. *J. Mater. Chem. B* 2, 428–436 (2014).

[85] Wen, J. et al. Diverse gatekeepers for mesoporous silica nanoparticle based drug delivery systems. *Chemical Society Reviews* (2017). doi:10.1039/c7cs00219j.

[86] Sun, Y. L., Zhou, Y., Li, Q. L. & Yang, Y. W. Enzyme-responsive supramolecular nanovalves crafted by mesoporous silica nanoparticles and choline-sulfonatocalix[4]arene [2]pseudorotaxanes for controlled cargo release. *Chem. Commun.* (2013). doi:10.1039/c3cc45216f.

[87] Zheng, Q. et al. A pH-responsive controlled release system using layered double hydroxide (LDH)-capped mesoporous silica nanoparticles. *J. Mater. Chem. B* 1, 1644–1648 (2013).

[88] Pishavar, E., Shafiei, M., Mehri, S., Ramezani, M. & Abnous, K. The effects of polyethylenimine/DNA nanoparticle on transcript levels of apoptosis-related genes. *Drug Chem. Toxicol.* 40, 406–409 (2017).

[89] Hunter, A. C. Molecular hurdles in polyfectin design and mechanistic background to polycation induced cytotoxicity.

*Advanced Drug Delivery Reviews* (2006). doi:10.1016/j.addr. 2006.09.008.

[90] Goyal, R., Bansal, R., Gandhi, R. P. & Gupta, K. C. Copolymers of covalently crosslinked linear and branched polyethylenimines as efficient nucleic acid carriers. *J. Biomed. Nanotechnol.* (2014). doi:10.1166/jbn.2014.1918.

[91] Goyal, R. et al. Linear PEI nanoparticles: Efficient pDNA/siRNA carriers in vitro and in vivo. *Nanomedicine Nanotechnology, Biol. Med.* 8, 167–175 (2012).

[92] M., W. C., L., R. M. & W, S. L. Systemic Circulation of Poly(L-lysine)/DNA Vectors is Influenced by Polycation Molecular Weight and Type of DNA: Differential Circulation in Mice and Rats and the Implications for Human Gene Therapy. *Blood* 97, 2221 (2001).

[93] Ambardekar, V. V. et al. The efficacy of nuclease-resistant Chol-siRNA in primary breast tumors following complexation with PLL-PEG(5K). *Biomaterials* (2013). doi:10.1016/j.biomaterials.2013. 03.021.

[94] Merdan, T., Kopeček, J. & Kissel, T. Prospects for cationic polymers in gene and oligonucleotide therapy against cancer. *Advanced Drug Delivery Reviews* (2002). doi:10.1016/S0169-409X(02)00046-7.

[95] Kogure, K. et al. Development of a non-viral multifunctional envelope-type nano device by a novel lipid film hydration method. *J. Control. Release* (2004). doi:10.1016/j.jconrel.2004.04.024.

[96] Nakamura, T., Akita, H., Yamada, Y., Hatakeyama, H. & Harashima, H. A multifunctional envelope-type nanodevice for use in nanomedicine: Concept and applications. *Acc. Chem. Res.* (2012). doi:10.1021/ar200254s.

[97] Cao, Y. et al. Aptamer-Conjugated Graphene Quantum Dots/Porphyrin Derivative Theranostic Agent for Intracellular Cancer-Related MicroRNA Detection and Fluorescence-Guided Photothermal/Photodynamic Synergetic Therapy. *ACS Appl. Mater. Interfaces* (2017). doi:10.1021/acsami.6b13150.

[98] Lin, G. et al. Quantum dots-siRNA nanoplexes for gene silencing in central nervous system tumor cells. *Front. Pharmacol.* (2017). doi:10.3389/fphar.2017.00182.

[99] Hardman, R. A toxicologic review of quantum dots: toxicity depends on physicochemical and environmental factors. *Environ. Health Perspect.* (2006). doi:10.1289/ehp.8284.

[100] Wahajuddin & Arora, S. Superparamagnetic iron oxide nanoparticles: Magnetic nanoplatforms as drug carriers. *International Journal of Nanomedicine* (2012). doi:10.2147/IJN.S30320.

[101] Namgung, R. et al. Hybrid superparamagnetic iron oxide nanoparticle-branched polyethylenimine magnetoplexes for gene transfection of vascular endothelial cells. *Biomaterials* (2010). doi:10.1016/j.biomaterials.2010.01.123.

[102] Hwang, J. H. et al. Heparin-coated superparamagnetic nanoparticle-mediated adeno-associated virus delivery for enhancing cellular transduction. *Int. J. Pharm.* (2011). doi:10.1016/j.ijpharm.2011. 10.019.

[103] Bates, K. & Kostarelos, K. Carbon nanotubes as vectors for gene therapy: Past achievements, present challenges and future goals. *Adv. Drug Deliv. Rev.* 65, 2023–2033 (2013).

[104] Singh, A., Trivedi, P. & Jain, N. K. Advances in siRNA delivery in cancer therapy. *Artificial Cells, Nanomedicine and Biotechnology* (2018). doi:10.1080/21691401.2017.1307210.

[105] Lee, J.-M., Yoon, T.-J. & Cho, Y.-S. Recent Developments in Nanoparticle-Based siRNA Delivery for Cancer Therapy. *Biomed Res. Int.* (2013). doi:10.1155/2013/782041.

[106] Kam, N. W. S., Liu, Z. & Dai, H. Functionalization of carbon nanotubes via cleavable disulfide bonds for efficient intracellular delivery of siRNA and potent gene silencing. *J. Am. Chem. Soc.* (2005). doi:10.1021/ja053962k.

[107] Li, H. et al. DOTAP functionalizing single-walled carbon nanotubes as non-viral vectors for efficient intracellular siRNA delivery. *Drug Deliv.* (2016). doi:10.3109/10717544.2014.919542.

[108] Lu, Q. et al. RNA polymer translocation with single-walled carbon nanotubes. *Nano Lett.* (2004). doi:10.1021/nl048326j.

[109] Bhatnagar, I., Venkatesan, J. & Kim, S. K. Polymer functionalized single walled carbon nanotubes mediated drug delivery of gliotoxin in cancer cells. *J. Biomed. Nanotechnol.* (2014). doi:10.1166/jbn. 2014.1677.

[110] Alhaddad, A. et al. Nanodiamond as a vector for siRNA delivery to Ewing sarcoma cells. *Small* (2011). doi:10.1002/smll.201101193.

[111] Bi, Y. Z., Zhang, Y. F., Cui, C. Y., Ren, L. L. & Jiang, X. Y. Gene-silencing effects of anti-survivin siRNA delivered by RGDV-functionalized nanodiamond carrier in the breast carcinoma cell line MCF-7. *Int. J. Nanomedicine* (2016). doi:10.2147/IJN.S117611.

[112] Akao, Y. et al. Microvesicle-mediated RNA molecule delivery system using monocytes/macrophages. *Mol. Ther.* 19, 395–399 (2011).

[113] Yang, T. et al. Delivery of Small Interfering RNA to Inhibit Vascular Endothelial Growth Factor in Zebrafish Using Natural Brain Endothelia Cell-Secreted Exosome Nanovesicles for the Treatment of Brain Cancer. *AAPS J.* (2016). doi:10.1208/s12248-016-0015-y.

[114] Lin, Y. et al. Exosome–Liposome Hybrid Nanoparticles Deliver CRISPR/Cas9 System in MSCs. *Adv. Sci.* 5, 1–9 (2018).

[115] Tibbitts, J., Canter, D., Graff, R., Smith, A. & Khawli, L. A. Key factors influencing ADME properties of therapeutic proteins: A need for ADME characterization in drug discovery and development. *mAbs* (2016). doi:10.1080/19420862.2015.1115937.

[116] Pottash, A. E., Kuffner, C., Noonan-Shueh, M. & Jay, S. M. Protein-based vehicles for biomimetic RNAi delivery. *J. Biol. Eng.* 13, 1–13 (2019).

[117] Yang, N. J. et al. Cytosolic delivery of siRNA by ultra-high affinity dsRNA binding proteins. *Nucleic Acids Res.* (2017). doi:10.1093/nar/gkx546.

[118] Pillé, J.-Y. et al. Intravenous Delivery of Anti-RhoA Small Interfering RNA Loaded in Nanoparticles of Chitosan in Mice:

Safety and Efficacy in Xenografted Aggressive Breast Cancer. *Hum. Gene Ther.* (2006). doi:10.1089/hum.2006.17.1019.

[119] Raftery, R., O'Brien, F. J. & Cryan, S. A. Chitosan for gene delivery and orthopedic tissue engineering applications. *Molecules* 18, 5611–5647 (2013).

[120] Jiang, H. L., Cui, P. F., Xie, R. L. & Cho, C. S. Chemical modification of chitosan for efficient gene therapy. in *Advances in Food and Nutrition Research* (2014). doi:10.1016/B978-0-12-800268-1.00006-8.

[121] Lohcharoenkal, W., Wang, L., Chen, Y. C. & Rojanasakul, Y. Protein nanoparticles as drug delivery carriers for cancer therapy. *Biomed Res. Int.* 2014, (2014).

[122] LeCher, J. C., Nowak, S. J. & McMurry, J. L. Breaking in and busting out: cell-penetrating peptides and the endosomal escape problem. *Biomol. Concepts* 8, 131–141 (2017).

[123] Guidotti, G., Brambilla, L. & Rossi, D. Cell-Penetrating Peptides: From Basic Research to Clinics. *Trends Pharmacol. Sci.* 38, 406–424 (2017).

[124] Ramakrishna, S. et al. Gene disruption by cell-penetrating peptide-mediated delivery of Cas9 protein and guide RNA. *Genome Res.* 24, 1020–1027 (2014).

[125] Axford, D. S., Morris, D. P. & McMurry, J. L. Cell penetrating peptide-mediated nuclear delivery of Cas9 to enhance the utility of CRISPR/Cas genome editing. *FASEB J.* (2017).

[126] Hernandez-Garcia, A. et al. Peptide–siRNA Supramolecular Particles for Neural Cell Transfection. *Adv. Sci.* (2019). doi:10.1002/advs.201801458.

[127] Tibbitt, M. W., Dahlman, J. E. & Langer, R. Emerging Frontiers in Drug Delivery. *J. Am. Chem. Soc.* 138, 704–717 (2016).

[128] Lee, H. et al. Molecularly self-assembled nucleic acid nanoparticles for targeted in vivo siRNA delivery. *Nat. Nanotechnol.* 7, 389–393 (2012).

[129] Bujold, K. E., Hsu, J. C. C. & Sleiman, H. F. Optimized DNA 'nanosuitcases' for Encapsulation and Conditional Release of siRNA. *J. Am. Chem. Soc.* 138, 14030–14038 (2016).

[130] Giljohann, D. A. et al. Spherical Nucleic Acid Nanoparticle Conjugates as an RNAi-Based Therapy for Glioblastoma. *Sci. Transl. Med.* 5, 209ra152-209ra152 (2013).

[131] Lee, J. B., Hong, J., Bonner, D. K., Poon, Z. & Hammond, P. T. Self-assembled RNA interference microsponges for efficient siRNA delivery. *Nat. Mater.* 11, 316–322 (2012).

[132] Roh, Y. H. et al. A Multi-RNAi Microsponge Platform for Simultaneous Controlled Delivery of Multiple Small Interfering RNAs. *Angew. Chemie - Int. Ed.* 55, 3347–3351 (2016).

[133] Qian, H. et al. Protecting microRNAs from RNase degradation with steric DNA nanostructures. *Chem. Sci.* 8, 1062–1067 (2017).

[134] Nahar, S., Nayak, A. K., Ghosh, A., Subudhi, U. & Maiti, S. Enhanced and synergistic downregulation of oncogenic miRNAs by self-assembled branched DNA. *Nanoscale* 10, 195–202 (2018).

[135] Sun, W. et al. Self-Assembled DNA Nanoclews for the Efficient Delivery of CRISPR-Cas9 for Genome Editing. *Angew. Chemie - Int. Ed.* 54, 12029–12033 (2015).

[136] Sun, W. et al. Cocoon-like self-degradable DNA nanoclew for anticancer drug delivery. *J. Am. Chem. Soc.* 136, 14722–14725 (2014).

[137] Xia, Y., Tian, J. & Chen, X. Effect of surface properties on liposomal siRNA delivery. *Biomaterials* (2016). doi:10.1016/j.biomaterials.2015.11.056.

[138] Maria Laura Immordino, Franco Dosio & Luigi Cattel. Stealth liposomes: review of the basic science, rationale, and clinical applications, existing and potential. *Int. J. Nanomedicine* (2006).

[139] Tu, Y. & Zhu, L. Lipids and Their Derivatives: By-Products Used as Essential Building Blocks for Modern Drug Delivery Systems. *Curr. Drug Targets* (2014). doi:10.2174/1389450115666140306154120.

[140] Peer, D. et al. Nanocarriers as an emerging platform for cancer therapy. *Nat. Nanotechnol.* 2, 751–760 (2007).

[141] Morrissey, D. V. et al. Potent and persistent in vivo anti-HBV activity of chemically modified siRNAs. *Nat. Biotechnol.* (2005). doi:10.1038/nbt1122.

[142] Chen, W., Li, H., Liu, Z. & Yuan, W. Lipopolyplex for therapeutic gene delivery and its application for the treatment of Parkinson's disease. *Frontiers in Aging Neuroscience* (2016). doi:10.3389/fnagi. 2016.00068.

[143] Aronsohn, A. I. & Hughes, J. A. Nuclear localization signal peptides enhance cationic liposome-mediated gene therapy. *J. Drug Target.* (1998). doi:10.3109/10611869808995871.

[144] Zuris, J. A. et al. Efficient Delivery of Genome-Editing Proteins In Vitro and In Vivo. *Nat Biotechnol* (2015). doi:10.1038/nbt.3081.

In: Gene Delivery
Editor: Vanessa Zimmer
ISBN: 978-1-53616-268-4
© 2019 Nova Science Publishers, Inc.

*Chapter 3*

# TARGETING TRANSPOSONS
# TO THE GENOME

*Jesse B. Owens*\*

Institute for Biogenesis Research, Anatomy,
Biochemistry and Physiology, John A. Burns School of Medicine,
University of Hawaii at Manoa, Honolulu, Hawaii, US

## ABSTRACT

Selfish DNA called transposons capable of cutting out and pasting into the host genome are active throughout the phylogenetic kingdoms. Researchers have repurposed natural transposons for use in delivering a gene-of-interest. This has enabled the study of a large and growing list of preclinical gene therapy applications. Hyperactive forms of the transposase enzyme have been developed, enabling high efficiencies of integration in mammalian cells. Recently, transposons have entered clinical trials for delivery of CARs for T-cell based immunotherapy. Several groups have attempted to control where insertions occur in the genome. By tethering DNA binding proteins to the transposase, the engineered fusion protein can be directed to a desired sequence and caused to integrate the transposon nearby. A targetable transposase could

---

\* Corrosponding Author Email: jbowens@hawaii.edu.

overcome drawbacks with virus-based approaches, including limited cargo capacity, host immune response, and the risk of insertional mutagenesis. However, major hurdles will need to be overcome before targetable transposition becomes available to the clinic. The specificity of these first-generation vectors is poor, which has stimulated new research focused on reducing off-target integration. The long-term goal is to generate a vector that exclusively integrates at the target sequence and nowhere else. This review discusses the past achievements and future challenges of this early-stage technology.

**Keywords:** targeted transposase, transposon, piggyBac, Sleeping Beauty, random insertion

# INTRODUCTION

When Dr. Barbara McClintock noted genetic material changing position in a maize chromosome in the 1950s, she launched the research that now holds promise in clinical trials to treat diseases such as acute lymphoblastic leukemia, non-Hodgkin lymphoma, and pancreatic cancer. She observed genetic elements that could "jump" and cause unstable mutations or restore gene functions when they transposed again. Dr. McClintock published this first discovery of transposons which was initially met with skepticism. The scientific community did not yet realize that she had glimpsed the beginnings of an innovative gene therapy. She was later awarded the Nobel Prize in Physiology or Medicine in 1983. Her discovery sparked research on "jumping genes", or transposons, that has led to novel studies on gene evolution, mutations, and disease development and treatment [1].

Transposons are genetic elements with the ability to change their position or "jump" in the genome. This jumping unit of DNA contains a transposase gene flanked by terminal inverted repeats (TIRs) containing transposase binding sites [2]. They make up about 50% of the human genome and are referred to as "genetic parasites" or "selfish DNA". Generally, the transposable elements "cut and paste". First, the transposase either randomly binds or binds to a specific DNA sequence. Then, the transposase "cuts" at the site and slices out the transposon DNA to "paste" at the target site. Researchers have replaced the contents of the transposon with their own custom DNA sequence. This allows researchers to deliver a gene of interest into a genome, a strategy that has opened the door to novel gene therapies [3]. There are a number of potential transposase candidates for use in genome editing, and they can be manipulated for greater safety and efficiency. The benefits and challenges of using the different transposases and transposon system designs are discussed in this review.

## Sleeping Beauty

After a long evolutionary sleep in the salmonid genome, Sleeping Beauty "awoke" to become the first transposase capable of transposition in vertebrate cells. Unlike the medieval princess, this Sleeping Beauty could relate more to Frankenstein since it was reconstructed from fossilized transposon sequences by researchers in 1997. Regardless, its generation and ability to "jump" in vertebrate cells enabled its use in gene therapy and raised new possibilities for future clinical studies [2].

## PiggyBac

Another transposase shown to function in eukaryotic cells is *piggyBac* (pB). PB was first discovered in 1983 in the cabbage looper moth *Trichoplusia ni*. It earned its name because its inserted DNA was carried "piggy-back" by *Baculovirus* [4]. In 1989, the sequence of pB was reported for the first time at 2,475 bp long and included encompassing TIRs with inserted fragments of DNA. In 1996, the transposase function of pB was confirmed with successful mobilization of a nonautonomous pB element. PB was found to be widespread, including in mammals as discovered with *piggyBat*. Its presence in a diversity of organisms led to the question, how do these parasitic elements affect the genomic evolution of their hosts [4, 5]? And, can they be manipulated to cause desired mutations and insertions in humans?

In 2005, pB was shown to be efficient in gene transfer in mice [6]. PB has a higher likelihood of hitting genes than SB but does not suffer from overproduction inhibition. Overall, pB has a higher efficiency of insertions, can sustain a large cargo size, and leaves no footprint when compared to SB [4, 5].

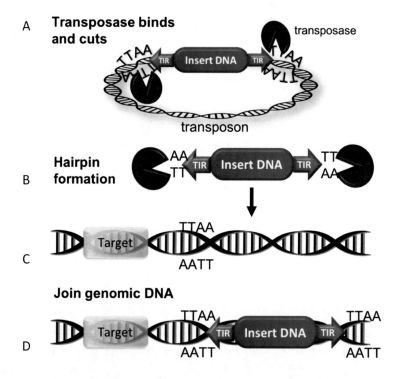

Figure 1. General insertion mechanism for pB transposase. A. The transposase binds and cuts out the insert DNA from the transposon. B. The transposases form a hairpin with the insert DNA. C. The transposase binds to the genomic DNA at TTAA sequences, ideally close to the sequence target of interest. D. The insert DNA is incorporated into the genomic DNA.

So, does pB "cut and paste" genomic sequences like one can with text on a computer? Sort of. Some transposases, like *piggyBac* (pB), do not require DNA synthesis. PB integrates at TTAA sites as shown in Figure 1. When the transposase is expressed, it binds to the inverted repeats of the transposon which inserts at a TTAA site. It nicks the DNA and frees the 3' hydroxyl groups at both ends of the transposon. This results in a hydrophilic attack of the flanking TTAA sequence, followed by hairpin formation. The transposon is consequently freed from the plasmid backbone which is repaired by ligation of complementary TTAA overhangs. The hairpin resolution of the transposon and the hydrophilic attack of gDNA by the 3' hydroxyl groups on the transposon allows for a four base pair (bp) cut in the gDNA and a transient double-stranded break with TTAA overhangs on both sides. There is no footprint mutation because when the transposase excises the transposon, it reforms a single TTAA, therefore it is seamless [4].

## DELIVERING GENES OF INTEREST

Research showed that as long as the optimal terminal length (311 bp for the 5' side and 235 bp for 3' side) was maintained, any DNA fragment could be inserted into the pB transposon. So, a researcher could insert a transcription unit if they wanted to express a gene in a host or disrupt gene function with a gene trap element. In addition, pB could fuse with DNA-binding domains, which would allow for site-directed transposition [5]. These factors are key benefits for its possible use in human genome therapy.

### Hyperactive Transposases

Transposases offer a promising platform for gene therapy but could be bottlenecked by gene transfer efficiency. An increased number of cells expressing the transgene could improve clinical applications. To mediate this obstacle, Mátés et al. generated a hyperactive transposase from Sleeping Beauty (SB) in 2009. After a large-scale genetic screen in mammalian cells they identified mutations that increased activity. They termed the improved transposase SB100X which had a 100-fold enhancement in efficiency [7].

Shortly after the creation of SB100X, Yusa et al. conducted a screen in yeast for mutations in pB that produced a rise in hyperactivity in 2011. They found 5/10,000 mutants with increased activity. They verified the activity of the mutants in mouse cells. Three-fourths of the mutants did not show hyperactivity in mice. Yusa et al. suggested that this could be due to species-specific factors positively or negatively affecting transposition, epigenetic modifications influencing frequency, or different reaction temperatures affecting catalytic activity or protein stability. Regardless, they identified seven mutations that increased activity in both yeast and mouse cells. They combined the seven active mutations to generate one hyperactive pB transposase (hyPBase). This new hyPBase has more than ten-fold higher rates of transposition than the WT pB and was even more efficient at integration than SB100X [8, 9].

Yusa et al. evaluated safety concerns for using hyPBase in gene therapy concerning both the transposon system and hyPBase itself. First, as with any gene delivery system, there is a risk of induced genome instability. They measured the excision-induced genomic alterations at the donor site and found a low frequency of 0.2%. Next, they did a genome-wide assessment of transposase-induced genome instability and did not observe large or complex events following hyPBase excision. In addition, they noted no obvious differences between cell lines exposed or not exposed to hyPBase and determined that the footprint was as low as the WT transposase at around 1%. Overall, they showed that hyPBase could mediate more efficient transposition without compromising genomic integrity [8].

## Applications of Transposon Systems

Transposon systems have already been proven effective in preclinical applications to combat disease. First, pB can be used for genome-wide insertional mutagenesis to identify genes involved in the expression of different phenotypes, including the development of different diseases. This method has been used for cancer gene discovery in mice. Rad et al. analyzed 63 hematopoietic tumors to reveal that pB is capable of genome-wide mutagenesis. They also uncovered previously unidentified genes involved in cancer [10]. In addition, Ikadai et al. used pB to screen for gametocyte-deficient mutants to combat the development of malaria [11].

In addition to their application in screening, transposases have been used in mice for genome editing. In 2000, Yant et al. successfully inserted DNA into mouse liver cells using the SB transposase and established DNA-mediated transposition as an efficient genetic tool for mammals [12]. In 2005, Ding et al. showed efficient transposition using the pB transposon carrying a marker gene in mice [6]. In 2009, Saridey et al. achieved long-term inducible gene expression *in vivo* in mice livers using the pB transposon system [13].

Transposon systems have been applied to human cells in addition to mice. For example, SB100X exhibited 35-50% stable gene transfer in human CD34+ cells. Transplantation to immunodeficient mice resulted in long-term engraftment and hematopoietic reconstitution. Additionally, expression of factor IX upon transposition of mouse liver *in vivo* was sustained for over a year [7].

Transposases can also be useful in cell-based regenerative medicine. Induced pluripotent stem cell (iPSC)-based regenerative medicine requires a nonmutated genome. Retroviral vectors can be tumorigenic and steps to avoid this result in low reprogramming efficiency. Yusa et al. developed a pB transposon-based approach that allows for a stable expression of transgenes that leaves no footprint mutations [14].

Transposases could also increase efficiency of other gene therapies. In 2012, Doherty et al. showed that hyPBase could increase gene transfer in human cells and *in vivo* in mice. Gene delivery increased two to three-fold long-term. HyPBase increases the number of cells expressing the transgene and is the most active system, including WT pB and SB100X, in human cells and *in vivo* in mice. Fusion of other proteins to hyPBase do not affect its activity, whereas SB100X loses significant activity with the addition of a site-specific DNA-binding domain [15, 16]. Its increased insertional efficiency and ability to be modified make hyPBase a prime clinical candidate to improve efficiency of gene therapies. HyPBase has been shown to increase the efficiency of T-cell modification by decreasing the time required to produce sufficient cells and delivering a greater quantity at one time. This could greatly improve the clinical response [9]. Finally, by injecting the transposon components into the pronucleus, hyPBase has enabled high rates of transgenic animal production [17].

Overall, transposon systems have allowed researchers to generate transgenic mice, manipulate embryonic stem cells, intentionally mutate the genome, generate induced pluripotent stem cells, and invoke long-term gene expression *in vivo* [9]. Specifically, pB has been shown to be effective in mammalian cells in various preclinical applications: gene discovery, multiplexed genome modification, animal transgenesis, gene transfer *in vivo* with long-term expression, and genetic modification of clinically-relevant cell types [4]. Preclinical studies of the use of SB in gene therapy include treatments for cancer, as well as muscle, neurological, pulmonary, hematologic, and metabolic disorders, and lysosomal storage and dermatologic diseases [5].

## Toward Use in Clinical Trials

In addition to preclinical applications, transposon systems are now being used in clinical trials. Transposons can genetically alter an isolated cell population that can then be transplanted into a patient [2]. This has many clinical applications. For example, investigators have replaced degenerated cells essential for vision with healthy cells generated with transposons to treat age-related macular degeneration (AMD) [18].

In addition, transposases can genetically alter T cells that can be used as immunotherapy for cancer. In 2010, Manuri et al. electroporated a pB system into human T cells to express a specific CAR that killed tumor targets. Genetically modified tumor-reactive T cells equipped with chimeric antigen receptors (CARs) are able to recognize surface molecules on tumor cells and exert antitumor effector functions. Transposons encoding CARs can be electroporated into T cells isolated from the patient's or donor's white blood cells which can then be expanded and administered to the patient [19].

Another potential use for engineered CAR T cells is in the treatment of pancreatic cancer. Pancreatic cancer, which is associated with the expression of mesothelin, has a poor prognosis due to the low efficacy of current treatments. He et al. developed mesothelin-targeting CAR T cells (mesoCAR T) using pB. They demonstrated suppressed tumor growth in the ASPC1 xenograft mice model, and observed more mesoCAR T cells differentiated into memory T cells after tumor remission with minimal lesions in major organs [20].

Transposon-delivered CARs have been effective against relapsed and refractory B cell malignancies. There was a sustained complete response rate of 80% against acute lymphoblastic leukemia and 50% against non-Hodgkin lymphoma. Designed CAR constructs without the IgG1 Pc region with different costimulatory and transmembrane domains could be efficiently expanded *ex vivo*, have shown potent activity against malignant B cells *in vitro,* and eradicate B cell xenografts *in vivo.* These studies are especially significant because it is less complex and more economical to generate CAR T cells via transposons than with viral vectors [21].

## Advantages of Using Transposons

Transposons have a greater cargo capacity than most viral vectors. PB can deliver inserts of over 100 kb [22], thereby allowing for the inclusion of multiple genes to be integrated in tandem, large endogenous promoters, or whole genes including introns and regulatory sequences [2]. Transposon vectors can be maintained as plasmid DNA, which is simple and inexpensive to make. Also, viral vectors are more expensive to generate and require regulation to ensure there are no replication-competent viruses. This is key for scaling up to clinical practice and becoming available at decentralized, local hospitals [2, 21].

Compared to retroviruses, SB and pB have less of a preference for inserting into genes. Transposons have a high efficiency of gene transfer which aids in the expansion of cells to clinically-viable numbers [4]. There is also a lower chance of immune response using DNA-based transposon systems compared to viruses that contain viral capsid proteins that the human immune system has evolved to fight.

## Random Insertion Risks

There is a risk with both transposon and viral integrating vectors of random insertion which could have negative consequences, such as the activation of proto-oncogenes, inactivation of tumor suppressor genes, or disruption of genes essential to organism fitness [23, 24]. One-fourth of the patients enrolled in two gene therapy trials for severe combined immune deficiency (SCID) developed T cell acute lymphoblastic leukemia. This was the result of vector-mediated insertional mutagenesis [25].

Targetable nucleases can be used to incorporate DNA but can result in undesired mutations. CRISPR nucleases primarily act passively; a double-strand break at the target site is followed by uncontrolled events mediated by host-cell proteins. The process of non-homologous end joining (NHEJ) introduces intentional mutations in dividing and nondividing cells and is crucial to CRISPR's efficient gene knock-out capabilities [26]. Homologous directed recombination (HDR) is an alternate process capable of incorporating donor DNA at a specific target sequence, albeit less efficiently [27]. NHEJ competes with HDR, usually resulting in a mutation without the desired insert [28]. HDR proteins are predominantly present during cell division [28, 29]. Therefore, HDR is not efficient in most tissues, which consist of rarely dividing or non-dividing cells. Recently, an analysis of CRISPR nuclease-edited mouse embryonic stem cells, mouse hematopoietic progenitors, and a human differentiated cell line uncovered unexpectedly high numbers of large deletions and chromosome rearrangements [30]. Previous studies may have underestimated the frequency of these multi-kilobase modifications because they are only detectable by long-read sequencing or long-range PCR genotyping.

Current gene therapy methods are associated with numerous safety issues. Targetable transposition offers great potential for ameliorating these risks and could be applied broadly to genome engineering studies [18]. With a targetable vector in hand, researchers could designate "safe harbors" in a genome and direct insertion toward those sites to avoid undesired mutations.

## SUCCESSES AND CHALLENGES IN TARGETED TRANSPOSITION

Transposases efficiently integrate large DNA sequences and can be delivered by non-viral, minimally immunogenic approaches [4, 31, 32]. Importantly, these technologies are capable of efficiently integrating into nondividing cells that compose most tissues. A method for user-defined transposon targeted integration with transient expression of the transposase would improve the safety and predictability of insertional therapies, while overcoming many of the limitations of viral and directed nuclease technologies. An intriguing hypothesis is that a transposase could be preferentially directed to specific DNA binding sequences, thereby performing the integration activity nearby [33]. Both indirect and direct fusions of a DNA binding domain (DBD) to the transposase have indeed allowed for targeted transposition. This area of research has greatly advanced over the past decade and novel developments are highlighted in the following studies and are summarized in Table 1.

**Table 1. Summary of research on targeting transposons**

| Trans-posase | Protein Interaction | Design Target | Methods to Check for Integration | On-Target Integr-ation? | Results Observations | Highlights | Ref |
|---|---|---|---|---|---|---|---|
| Tn7 | TnsABC+D system | *glms* safe site | NA | NA | Highly specific for safe site in bacteria | Naturally occurring transposon | [34, 35] |
| ISY100 | Fuse Zif268 zinc finger | Promoter region of *HIV-1* (variable spacing) | Plasmid to plasmid, digest, E. coli | Y | ~50% of 24 transposition products on-target in bacteria | Optimized target site spacing and linker length | [36] |
| IS30 | Fuse cI repressor of bacteriopha-ge λ | λ operator O$_R$ | Plasmid to plasmid, direct sequencing, E. coli | Y | 30/53 insertions close to target and 1/53 directly on-target | High number of off-target insertions, targeting achievable in living vertebrate cells | [37] |
| | Fuse Gli1 | Gli1 binding site | Nested genomic PCR, zebrafish | Y | Single confirmed transposition event | | |
| Mos1 pB | Fuse Gal4 | UAS | Plasmid to plasmid, direct sequencing, insect embryos | Y | N- and C-terminal fusions integrate closer to the target than native pB or Mos1 | Highly specific, site-directed integration in insect cells | [16, 38] |
| SB | Fuse Gal4 | UAS | Plasmid to plasmid, direct sequencing, human cells | Y | More integrations with fusion and with an imperfect target site | Transient interaction may be preferable over stable docking | [39] |
| | Fuse E2C zinc finger | *erbB-2* | | Y | | | |
| SB | Bridge protein N-57 fused to TetR | Artificially inserted tetracycline response element | Nested PCR, isolation of clones, human cells | Y | 12 genomic insertions sites with 10% recovered in isolated clones | Used a bridging protein that binds to SB, not fused. | [40] |
| pB | Fuse CHK2 zinc finger | Promoter of checkpoint kinase-2 | Plasmid to plasmid, nested genomic PCR, human cells | Y | Double the insertions recovered for fused transposase vs. native to introduced genomic sequences | More targeted insertions with fused constructs | [41] |

|  | Design |  |  |  | Results |  |  |
|---|---|---|---|---|---|---|---|
| Trans-posase | Protein Interaction | Target | Methods to Check for Integration | On-Target Integr-ation? | Observations | Highlights | Ref |
| pB | Fuse N-terminal Gal4 | UAS | Plasmid to plasmid, nested genomic PCR, human cells | Y | N-terminal fusion facilitated targeting to introduced and endogenous sequences. C-terminal only targeted introduced | First example of targeting endogenous sequences | [15] |
|  | Fuse C-terminal Gal4 |  |  | Y/N |  |  |  |
| SB100X | Fuse ZF-B zinc finger | Human LINE1 element | LAM PCR, human cells | Y | 1.36% targeted insertions when fused with ZF-B and 0.82% when bound to N57 fused to ZF-B, compared to 0.34% SB100X alone | Fusion increased targeting to endogenous genome by 4-fold | [42] |
|  | Bridge protein N-57 fused to ZF-B |  |  | Y |  |  |  |
| Native SB | Bridge protein N-57 fused to Rep | RRS | LAM PCR, human cells | Y | 0.339% when bound to N57 fused to Rep, compared to 0.184% SB alone | Use of a bridging protein increased targeting by 2.7-fold | [43] |
| pB | Fuse SPI | SPI binding site | Inverse PCR | Y | 6,373 potential binding sites recovered, 83% confirmed SP1 binding site within 250 bp | "Calling card" to predict DBD binding sequences | [44] |
| hyPBa-se | Fuse TALC1 TALE | CCR5 | Nested PCR, isolation of clones, human cells | Y | Recovered 0.014% and 0.010% isolates for first and second strategies. Isolated clones with single targeted insertions. | First targeting of single, custom, endogenous site. Directed toward clinically-relevant CCR5 sequence | [45] |
|  | Fuse Gal4 + Bridge protein Gal4 fused to TALC1 |  |  | Y |  |  |  |
|  | Fuse TALR1 + Bridge protein TALR1 fused to TALC1 |  |  | Y |  |  |  |
|  | Fuse TALC1 and Gal4 |  |  | N |  |  |  |
|  | Fuse Gal4 and TALC1 |  |  | Y |  |  |  |

# Table 1. (Continued)

| Design | | | | Results | | | |
|---|---|---|---|---|---|---|---|
| Trans-posase | Protein Interaction | Target | Methods to Check for Integration | On-Target Integr-ation? | Observations | Highlights | Ref |
| pB | Fuse HPRT zinc finger | HPRT | Selection of on-target cells using 6-TG. | Y | Targeted integration was 0.01% for pB, 0.45% for ZFP-pB, 0.97% for TALC-pB, and none for Cas9-pB. | Comparison between different DBDs. Showed that RNA-guided transposition is difficult in human cells. | [46] |
| | Fuse HPRT TALE | | | Y | | | |
| | Fuse SpCas9 with HPRT gRNAs | | Nested genomic PCR, deep sequencing, human cells. | N | | | |
| TnsB, TnsC, TniQ | Unfused ShCas12k | 48 sites in the E. coli genome | Nested PCR, deep sequencing, E. coli | Y | >50% of insertions on-target. Insertions occurred at a predictable distance from gRNA target site | Re-programmed native CRISPR/Transposase to target E. coli with high specificity | [47] |
| | Unfused AcCas12k | | | Y | | | |
| TnsA, TnsB, TnsC, TniQ | Unfused Cas6, Cas7, Cas8-Cas5 | 16 sites in the E. coli genome | Nested PCR, deep sequencing, E. coli | Y | 95% targeting specificity for 16 guide RNAs | | [48] |
| pB excise-on only | Fuse ROSA26 or GULOP zinc fingers | ROSA26 and | LAM PCR, human cells | N | No targeted integration | Additional modification-ns needed for "site required" targetable transposase | [49] |
| pB partial excise-on only | | GULOP safe harbors | | N | | | |
| hyPBa-se | Fuse dCas9-HF1 with CCR5 gRNAs | CCR5 | Nested PCR, isolation of clones, human cells | Y | Total of 17 independent insertion sites recovered, 0.06% of total cells contained targeted insertions. Isolated clones with single targeted insertions. | RNA-guided transposition in human cells is possible | [50] |
| hyPBa-se partial excise-on only (H2) | | | | Y | | | |
| hyPBa-se excise-on only (H3) | | | | Y | | | |

## Safe site Insertion Gives Direction to Targeting Studies

Researchers can employ transposases to insert DNA at a "safe site". In nature, the bacterial Tn7 transposon directs its insertion several bps downstream of the glutamine synthetase (*glmS*) gene. Insertions to this locus are advantageous because neither *glmS* nor any other essential bacterial genes are disrupted. The Tn7 transposon encodes TnsD which contains a DBD that recognizes the *glmS* locus called attTn7. Through interaction with other Tn7 encoded proteins, TnsD directs transposon insertion to this "safe" site (Figure 2) [34, 35]. These studies have set the stage for future work focusing on targeting safe or clinically-relevant sites.

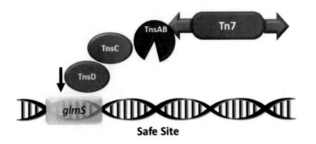

Figure 2. Tn7 transposon insertion targeting mechanism with TnsABC+D. TnsD binds to the target site (*glmS*) and recruits TnsC. TnC forms a platform to receive the TnsAB transposase bound to the Tn7 transposon which is inserted downstream of *glmS* at a 'safe site'.

## Optimal Target Site Spacing and Linker Increases Targeted Insertions

Using a bacterial system, the ISY100 transposase was fused to the mouse transcription factor Zif268 and two engineered Zif268 variants that bind specifically to the promoter region of HIV-1 (Figure 3) [36]. Highly specific targeted transposition was achieved using an assay in which a transposon was transferred from a donor plasmid to a recipient plasmid harboring a target sequence. Most insertions were located only 7–17 bp from the Zif268 binding site. In addition, multiple target sequences were tested by incrementally increasing the spacing between the zinc finger and target "TA" dinucleotide insertion site. Targeted insertions occurred 19-times more frequently when using optimal target site spacing and using the optimal linker between the DBD and transposase. [36].

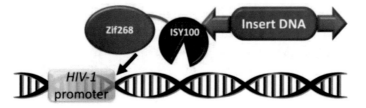

Figure 3. The Zif268 zinc finger DBD directs ISY00 transposase to target the *HIV-1* promoter site.

## Targeted Insertion Possible in Living Eukaryotic Cells

IS30 was the first prokaryotic transposon shown to target in vertebrate cells [37]. This *E. coli* transposase was first fused to the cI repressor of bacteriophage λ and the activity of the chimeric protein was examined in an *E. coli* two-component system consisting of a donor plasmid containing the transposon and fusion transposase as well as a recipient plasmid carrying the λ operator ($O_R$) (Figure 4A). 30/53 of recovered insertions occurred within 400bp of the $O_R$. IS30 was also fused to the Gli1 DBD and was expressed in zebrafish embryos (Figure 4B). A plasmid to plasmid gene trap method in which a successful targeting event positioned a promoterless GFP downstream of a promotor found on a recipient plasmid was used to detect insertions near the Gli1 9-bp recognition sequence. 3/104 embryos expressed GFP when injected with the complete assay but none were detected without either the fusion transposase or target sequence. In this study, only a single transposition event was recovered at 36 bp from the target sequence. Unexpectedly, a much higher number of insertions due to illegitimate recombination were caused by the fusion of Gli1 to IS30 [37]. Overall, there was a high number of off-target insertions in this study but the ability of this mechanism to target specific sequences in living cells opened the door to the possibility of targeting with transposases in human gene therapy.

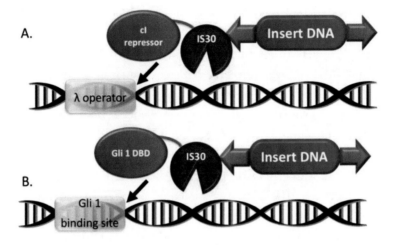

Figure 4. Mechanism of IS30 fused to a cI repressor (A) or Gli1 DBD (B) for targeted transposition in vertebrate cells.

## Site-Directed Integration in Insect Cells

In ordered to dial in on efficiency and increase the on-target percentage of targeted transposases, numerous fusion constructs and methods have been employed. For example, a plasmid-based transposition assay performed in *Aedes aegypti* embryos was used to analyze the efficiency of Gal4/Mos1 and Gal4/pB chimeric transposases [38]. The Gal4 DBD binds to its upstream activating sequence (UAS) with high affinity. It was fused to the Mos1 transposase isolated from *Drosophila mauritiana* and the pB transposase from the cabbage looper moth, *Trichoplusia ni* (Figure 5A and B). Transposase expression plasmids were coinjected with a recipient plasmid harboring five UAS copies and a donor plasmid with a transposon containing the kanamycin resistance gene. Successful transfer of the transposon resulted in a double antibiotic recipient plasmid that could be isolated and sequenced upon transformation of *E. coli*. The transpositional activity of the Gal4/Mos1 transposase was 12.7-fold higher compared to controls without the UAS target sequence. This was likely due to the increased availability of the recipient plasmid adjacent to Mos1 and not from an increase in enzymatic activity of the transposase. Remarkably, 96% of the Gal4/Mos1 recovered products were identical, with integration occurring at the same TA site 954 bp from the target sequence. Of these insertions, 98% were in the same 5' to 3' orientation. PB, which inserts at TTAA sites, also showed a marked increase in transpositional activity (11.6-fold higher) with 67% of the integrations occurring at a single TTAA site. Among the inserts located at the single site 912 bp from the UAS, 80% were in the same 5' to 3' orientation. Overall, Gal4 resulted in the greatest transposase activity and had 96% and 67% on-target insertions when fused with Mos1 and pB, respectively. In a separate publication, the genomic integration efficiency was compared between pB, the SB version SB11, Tol2, and Mos1 in four mammalian cell lines. Although only integration efficiency and not specificity was tested, addition of Gal4 to pB resulted in ~92% activity of the fusion transposase, whereas all other fusions abolished activity [16].

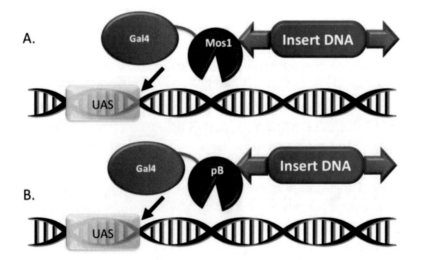

Figure 5. Site-directed integration in insect genomes. Gal4 fused to Mos1 (A) and pB (B) increased the on-target percentage of insertions.

## Transient Interaction Preferable over Stable Docking

The consensus sequence for an ancient fish transposon that had become inactivated through evolution was generated by comparing sequences between salmonid subfamily members. By identifying and reverting these inactivating mutations, SB was "awoken" and was shown to be fully active in human cells [51]. In an attempt to target insertions in human cells, the SB transposase was fused to a variable-length linker to either end of Gal4 or E2C zinc-finger DBDs as shown in Figure 6A and B [39]. C-terminus fusions abolished activity but SB transposases with N-terminal additions retained ~10% activity. To test for targeting ability, a similar plasmid-based assay was used as Maragathavally et al., described above [38]. Three plasmids were cotransfected, including a donor encoding bleomycin in the transposon, a helper encoding the chimeric SB transposase, and an ampicillin resistant recipient plasmid containing five tandem repeats of the DBD recognition site. Compared to native SB, the Gal4 and E2C fusion proteins directed 11- and 8-fold more integration events into the 443 bp targeting window, respectively.

Interestingly, a control target site with an imperfect binding sequence improved targeting instead of abolishing it. The mutant was expected to support base pair contacts with only half of E2C. The authors suggest that stable docking may constrain the transposase and that a transient interaction may be preferable. Also, the zinc-finger E2C binds a unique sequence on human chromosome 17. Unfortunately, following expression of E2C/SB, no genomic insertions were recovered from the E2C locus. The reason for this was unclear but may be due to an unfavorable surrounding sequence to the E2C site.

While the transposases fused to proteins were more active (had more integration events) than the native transposase, the on-target percentages for this study ranged from 17.8-25%. Since the imperfect binding sequence improved targeting, this research suggests that transient interaction is preferable over stable docking.

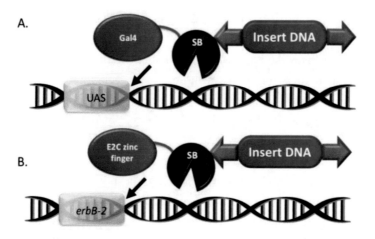

Figure 6. Targeting mechanism using Gal4 (A) and the E2C zinc finger (B) fusions to SB. An imperfect binding sequence of *erbB-2* improved binding.

## Targeting with a Bridge Construct to Direct a Transposase

A different group attempted to target SB insertions by generating a fusion construct with the addition of a bridge protein. This group first attempted a direct fusion of a DBD to SB [40] but a number of proteins lost all activity, similar to what had been seen previously [16]. A fusion to the Jazz zinc finger that remained 10% active failed at targeted transposition.

Alternatively, instead of a direct fusion, the transposon DNA itself was tethered to the target site by a bridge protein consisting of two DBDs. The transposon contained the LexA operon binding sequence which interacts with the LexA DBD found on the double DBD bridging protein. The tetracycline response element target site was randomly integrated chromosomally into cultured human cells. The second DBD on the double DBD bridging protein, TetR, binds to the tetracycline response element specifically, thus allowing for the targeting of the transposon (Figure 7A)[40]. Native unfused SB, the double DBD protein, and a donor plasmid were cotransfected into Hela cells and genomic DNA was screened for insertions by PCR. Two insertions, each in a different 5' to 3' orientation, were recovered 44 and 48 bp downstream of the target region.

A third strategy was employed in which the double DBD protein consisted of TetR and a protein domain that interacts with the SB transposase called N-57. Instead of targeting the transposon DNA, this configuration targeted the SB transposase protein to the tetracycline response element (Figure 7B). In total, 12 insertion sites were recovered from 6 polyclonal transfections, from which 3 sites were hit multiple times independently. A hotspot 44 bp downstream of the target site was hit 8 times and targeted insertion occurred within a 2.6-kilobase window around the tetracycline response element. Impressively, 10% of isolated single clones contained at least one targeted insertion, suggesting the efficiency of using a bridge construct to direct transposition.

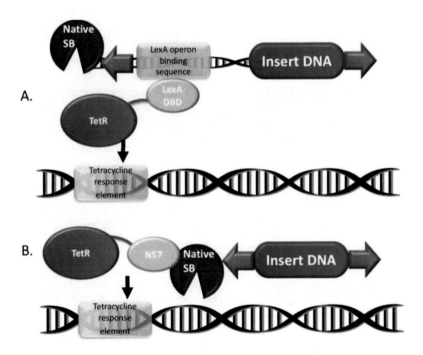

Figure 7. A. A bridge protein consisting of two DBDs tethers the transposon DNA to the target tetracycline response element. B. N57 interacts with the SB transposase. A bridge protein consisting of N57 and TetR tethers the transposase enzyme to the tetracycline response element.

## Double the Amount of Insertions Recovered with Fused Transposases

Another study compared the activity and on-target percentage of using fused transposase constructs vs. native transposases. A chimeric transposase, in this case the pB transposase fused to a custom zinc finger engineered to bind the promoter of checkpoint kinase-2 (CHK2), was tested for targeted transposition in human cells (Figure 8) [41]. A plasmid to plasmid assay was designed to have TTAA-rich sequences flanking the 18 bp zinc finger target site or a scrambled 18-bp target sequence. This recipient plasmid contained TTAA sites every 25 bp for the first 100 bp in either direction and then TTAA sites every 50 bp for the next 150 bp. The number of recovered double antibiotic colonies from cotransfections using the CHK2/pB fusion increased 3.5 fold compared to native pB. This was likely due to the increased availability of the recipient plasmid to the transposase due to the tethering by the zinc finger.

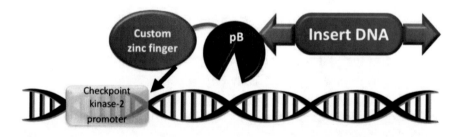

Figure 8. Targeting with fusion of custom zinc finger to pB resulted in more insertions on or near the checkpoint kinase-2 promoter target than using unfused, native pB.

Importantly, statistically more insertions occurred near the CHK2 binding site with 74% of CHK2/pB insertions recovered within 500 bp compared to only 50% for native pB. Unfortunately, site-directed integration to the endogenous genomic CHK2 site did not occur. This may have been the result of the scarcity of TTAA sites surrounding the binding site. The same TTAA-rich target sequence from the plasmid to plasmid assay was integrated into the genome and transpositional targeting was assayed using either native pB or the CHK2 chimeric pB, but results were confounded due to the efficient recovery of negative control native pB insertions into the target region. The mechanism for why an unfused pB would selectively target this handful of TTAA sites given the millions of potential alternate TTAA sites throughout the human genome was not elucidated. Even so, there were double the targeted insertions recovered for CHK2/pB than for native pB.

## Using DNA Binding Domains to Target Endogenous Sequences in the Genome

In a parallel set of experiments, the ability of both N- and C-terminal fusions of the Gal4 DBD to the pB transposase was demonstrated to target plasmids and the genome in human cells (Figure 9A and B)[15]. Using a similar plasmid to plasmid approach as above, five copies of the Gal4 UAS binding sequence were flanked on either side by 65 TTAA sites spaced 10 bp apart. The intention was to analyze the optimal distance of integration from the UAS that is likely dictated by protein tension or linker length. The amount of total integrations was over three times that of native pB control. This was similar to what had been reported previously and was likely due to Gal4 joining the pB transposase and recipient plasmid together via UAS target binding. By analyzing the distance of integration sites from the UAS, it was found that 87% of N-terminal Gal4/pB insertions were located within a region 800 bp up- or downstream of the UAS, whereas only 59% of the pB control insertions fell within this region. Similarly, the C-terminal fused pB also had a significant ability (77%) to integrate near the UAS site. Interestingly, 39% of N-terminal Gal4/pB insertions were found within 250 bp upstream of the UAS site compared to only 9% for native pB.

The genomic targeting scenario was analyzed by loading the same UAS and TTAA-rich regions from the plasmid assay onto the SB transposase and randomly integrating these target sites into the genome. The researchers first determined that each cell contained one to two of these target transposons. They retransfected these cells with N- or C-terminal or native pB and ran nested genomic PCR to identify insertions near the UAS. They did not obtain PCR products from native pB transfections or from any transfections of a control cell line which was absent for the UAS. For the chimeric pB transfections, they recovered 49 unique insertions, in which 96% localized within 800bp of the UAS. Seven sites were targeted three or more times and eight loci shared integrations from both fusion constructs.

The Gal4 DBD recognizes a specific 6bp binding site found 56,898 times in the human genome. Using nr-LAM PCR, the investigators identified over 7,000 genomic insertion sites for the pB constructs and compared the distances from these insertions from endogenous Gal4 recognition sites. They only detected increased targeting for the N-terminal fusion in which insertions landed within 0.8 kb of a Gal4 site 23% of the time compared to 5% of the time for control. Additionally, 32% of insertions landed within 1.8 kb compared to 8% for control. The difference of 24% represents the percentage of targeted integrations due to the presence of the Gal4 DBD [15]. Successful targeting of endogenous sequences implies that alternative, and potentially more specific, DBDs could be used to target the genome.

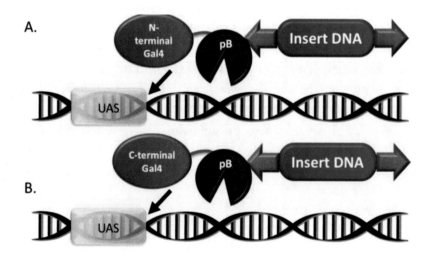

Figure 9. N-terminal (A) and C-terminal (B) Gal4 fusions to pB increased on-target insertions.

## Fusion Increases Targeting Efficiency in Endogenous Loci

In order to assay targeting to endogenous loci using SB, a custom zinc finger called ZF-B was generated and was incorporated into two targeting designs [42]. The target was an 18-bp sequence located on human LINE1 elements found about 12,000 times in the genome. About 17% of the human genome consists of LINE1 elements but only a fraction contains this zinc finger binding sequence. These sites were chosen because of their high frequency, low GC-content, and location in gene-poor regions. Ideally, insertions into LINE1 elements would disrupt genes less often than random. The first design was a direct fusion of ZF-B to SB (Figure 10A). The optimal configuration included an improved linker over previous attempts and the use of the hyperactive SB100X transposase which resulted in 40% activity compared to unfused SB100X. This activity is substantially higher than earlier reported fusion proteins [16, 39]. The second design included a bridging protein consisting of the ZF-B zinc finger fused to the N57 protein that binds SB (Figure 10B). This design was analogous to an earlier experiment used to direct insertions near the tetracycline response element [40]. Unfused SB was mixed with the bridging N57/ZF-B protein to tether SB to the target site. Genomic insertion sites were recovered using linear amplification mediated (LAM) PCR following transfection of unfused SB100X transposase, the ZF-B/SB100X fusion protein and the N57/ZF-B bridging protein mixed with unfused SB100X transposase. Insertions were significantly enriched in LINE1 elements containing the ZF-B binding site.

Only 7% of unfused SB100X insertions were recovered in these LINE1 elements compared to 11% and 13% for the ZF-B/SB100X fusion and N57/ZF-B mixed with SB100X designs, respectively. Also, insertions occurring within a 400-bp window surrounding ZF-B binding sites were counted. 0.34% of insertions were recovered using unfused SB compared to 1.36% for the ZF-B/SB100X fusion protein and 0.82% for N57/ZF-B mixed with unfused SB100X transposase. For the ZF-B/SB100X fusion, the difference of about 1% represents insertions due to the fused ZF-B

DBD. Targeted insertions were four-times more likely to occur using the fusion transposase strategy than using unfused SB [42].

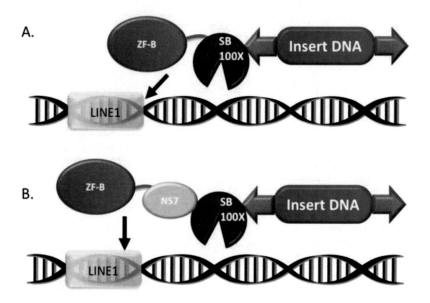

Figure 10. SB100X fused to ZF-B (A) and guided by N57 fused to ZF-B (B) targeted the human LINE1 element.

## Analyzing the Efficiency of Using Bridging Proteins

The same group also attempted to use a bridging protein consisting of the N57 domain that interacts with SB and the Rep DBD from AAV that binds to an endogenous human sequence called RRS found on ch19 within the phosphatase 1 regulatory subunit 12C (PPP1R12C) gene [43]. This gene encodes a protein with a function that is not clearly delineated. Cotransfecting unmodified SB with this N57/Rep protein is expected to redirect the SB transposase near RRS sites in the genome (Figure 11). The N57 domain, in addition to binding to the SB protein, also binds to the SB inverted repeats on the transposon. Therefore, the bridging protein N57/Rep could theoretically tether the transposon DNA to the RRS sequence as well. To enhance this interaction, additional N57 binding sites were added to the transposon. Using LAM-PCR to identify insertions occurring within 5kb of RRS-like sites (RRS consensus sequence with up to 2 mismatches), a bias in targeted integration was found. Specifically, 0.339% of insertions using the N57/Rep + native SB were recovered compared to 0.184% of insertions for native SB alone, representing a 2.7-fold enrichment. A similar experiment using a direct fusion of Rep to pB did not result in robust targeting to RRS sites. However, a direct fusion of Rep to Tol2, although < 10% efficient at total integration than native Tol2, resulted in ~4-fold increase of Tol2 transposon integrations into 5-kb and 10-kb windows around RRS-like sites. Overall, use of a bridge protein increased recovery of on-target insertions.

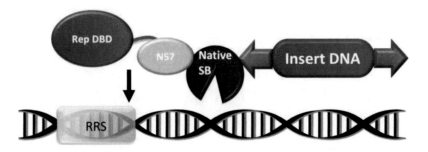

Figure 11. N57 fused to the Rep protein directs native SB to the RRS site in the genome.

## Calling Cards to Identify DBDs Relevant to Diseases

An interesting application for endogenous transpositional targeting, in addition to targeting safe sites, are "calling cards" which have been used to predict DBD binding sequences in the human genome [44]. The method relies on targeting pB insertions by fusing a C-terminally bound DBD of interest to the transposase. Resulting transposon insertion sites are identified by inverse PCR in which a common cutting enzyme is used to cut at a known site within the transposon and cut an unknown site in the genome. These fragments are self-ligated and the resulting circular DNA can be used as a template for the PCR amplification of the transposon-genome junction. These products are then deep sequenced and mapped. For a given dataset, there are a mix of both targeted insertions near the DBD recognition sequences as well as off-target insertions. However, the chance that two off-target insertions occur near one another is low. Therefore, by focusing on insertion clusters in which multiple insertions occur near one another or at the same TTAA site, a great number of off-target insertions can be removed from the analysis. Indeed, a calling card experiment analyzing insertions of the SP1 transcription factor fused to pB gave rise to 6,373 significant clusters, and remarkably, 83% of these contained an SP1 binding site within 250 bp (Figure 12). These results were reproducible, as top clusters identified by two replicates showed a 72% overlap. The traditional method for identifying DBD recognition sequences, CHIP-seq, was used to verify the calling cards strategy. SP1-directed calling card clusters were located with 250bp of ChIP-seq peaks 80% of the time compared to only 5% for native pB control. This method and modifications of this strategy can be employed to elucidate DBD binding sequences in the human genome with roles in disease development.

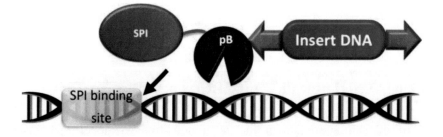

Figure 12. Transposon targeting mechanism to leave "calling cards". Sequencing out from the inserted transposons identifies SP1 binding sites in the genome.

## Working toward Efficient on-Target Transposition to Desired Sequences

A major goal for targetable transposition is to provide researchers with a means for inserting at a single genomic location that can be chosen. The proof of concept experiments demonstrating targeting to endogenous recognition sequences such as Gal4, Rep, and SP1 used DBDs that cannot be programmed to a desired sequence. Furthermore, these DBDs each recognize short motifs and bind thousands of sites throughout the genome. On the other hand, TALE DBDs can be generated to bind to > 18 bp motifs with sufficient specificity to target a single custom sequence in the genome.

One group tested whether TALEs could direct transposon insertion by generating a variety of constructs using combinations of both plasmid-DNA and transposase-protein localization to a unique target sequence [45]. A custom TALE, termed TALC1, was designed to bind the first intron of the C-C chemokine receptor type 5 (*CCR5*) gene. This locus has been identified as a "safe harbor" due to its distance from cancer-related genes and because knockout of *CCR5* not only is asymptomatic in humans, but also has the added benefit of providing resistance to R5 tropic strains of HIV-1 [52].

They developed five total strategies. The first was analogous to the previous Gal4 direct fusion to pB that, instead, fused TALC1 to the N-terminal of pB (Figure 13A). The second strategy used a bridging protein to tether the plasmid DNA containing the transposon to the TALC1 site Figure 13B). This was achieved by fusing Gal4 to TALC1 to make a double-DBD protein. To promote binding to the plasmid, UAS sequences were added to the plasmid backbone. Additionally, a second Gal4 was fused to the pB transposase so that pB would localize near the transposon which, in turn, was localized near the genomic target site. The interaction of a TALE with its target sequence is more specific than with Gal4 due to the TALE's longer binding motif. Therefore, the third strategy was similar to the second except that a TALE was used in place of Gal4 (Figure 13C). The fourth and fifth strategies fused the Gal4/TALC1/pB and TALC1/Gal4/pB together, respectively (Figure 13D). These single protein chains directly fused pB to the TALE and included Gal4 which, in turn, could bind the plasmid. Therefore, both transposase enzyme and transposon DNA could theoretically be targeted to the TALC1 site.

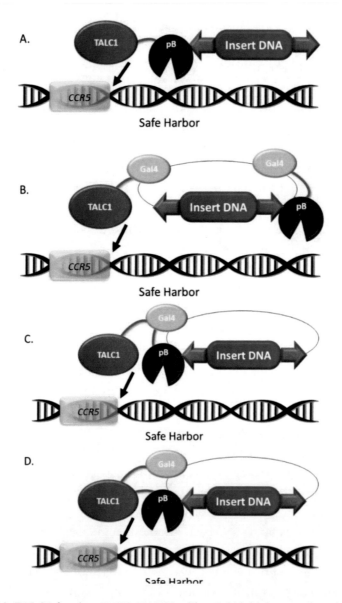

Figure 13. TALC1 fused to pB (A), TALC1 with a Gal4 bridging protein guiding pB with an additional fused Gal4 (B), TALC1 with a TALE bridging protein guiding pB with an additional fused Gal4 (C) targeted clinically-relevant site *CCR5*. TALC1 with a TALE or Gal4 bridging protein guiding pB with an additional fused Gal4 and an addition pB fused to TALC1 to target TALC1 and TALC2 recognition sequences (D).

Following transfection into HEK293 cells, a total of 14 insertion sites were recovered by nested PCR from the *CCR5* locus. Targeted insertion events were identified from all but the fourth strategy and native pB also did not target *CCR5*. One site, located 24 bp upstream of the TALC1 recognition sequence, was recovered from four independent transfections, two of which used the direct fusion. Additionally, 9 of the 14 insertion sites were located within 250 bp of the TALC1 sequence.

The overall targeting efficiency of the first and second strategy was determined by screening small pools of cells and was calculated to be 0.014 and 0.10%, respectively. In order to be able to identify and expand these rare cells, a two-step PCR screen was used. By screening a single 96-well dish containing ~56 colonies per well, a positive well was identified. Next, this well was single-cell sorted and approximately 1/56 single-cell wells was found to contain correctly targeted cells. Therefore, following the pre-screen, ~2% of cells were targeted. In this experiment, they recovered five of these correct colonies and used a copy-number assay to confirm that these cells contained a single insertion. Because the location of this insertion had been determined by nested PCR to be 24bp from the TALC1 site in *CCR5*, they reasoned that no other off-target insertions were present in these cells. Long-term culture did not result in transgene silencing, confirming that *CCR5* was a suitable safe harbor target [45].

## Steps toward Engineering Targeted pB

In 2017, Luo et al. evaluated three different targeting systems fused to the pB transposase. They compared different proteins fused to the pB transposon in order to target the *hypoxanthine phosphoribosyltransferase* (*HPRT*) gene. They evaluated three engineered zinc finger proteins (ZFP), four transcription activator-like effector proteins (TALE), and five different guide RNAs (gRNAs) with a non-cutting CRISPR-associated protein 9 (SpCas9).

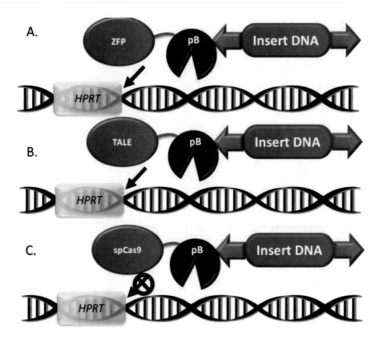

Figure 14. Fusion of pB to ZFP (A) and TALE (B) resulted in targeted integration at HPRT. No targeted integration occurred with the fusion of spCas9 (C).

First, they integrated the pB transposon into HT-1080 cells with one copy of *HPRT*. If the gene remains active, addition of 6-thioguanine (6-TG) is toxic which indicates that there is no targeted integration.

They evaluated the chimeric transposases for expression, transposition activity, chromatin immunoprecipitation, and targeted knock-out (KO) of *HPRT*. One ZFP-pB (Figure14A) and one TALE-pB (Figure 14B) successfully targeted *HPRT* but, like previous studies, contained off-target integrations as well. TALE-pB had the most recovered colonies (almost 200 per 1x10$^6$ cells), with ZFP-pB close behind (almost 100 per 1x10$^6$ cells).

To map integration into *HPRT*, Luo et al. used PCR with *HPRT*-specific primers, then sequenced. They also recovered insertions using deep sequencing. They found four ZFP-pB insertions and three TALC-pB insertions using nested PCR. Deep sequencing revealed three frequently hit hotspots for ZFP-pB and five for TALC-pB. One of the hotspots for ZFP-PB was found within 200 bp of the binding site, whereas TAL2-PB hotspots were more dispersed throughout *HPRT*. Both chimeric transposase fusion proteins integrated at a hotspot near the transcription start site. The overall percent of targeted transposon integration out of total integration was 0.01% for pB, 0.45% for ZFP-pB, 0.97% for TALC-pB, and none for Cas9-pB. They calculated a range of 0.019% for TALE targeting, similar to Owens et al. above [45, 46].

Unfortunately, they were not able to employ gRNA-mediated targeted transposition into a user-defined chromosomal locus using SpCas9-pB (Figure 14C). They used multiple gRNAs together with dCas9-pB but found no improvement. Luo et al. suggested that SpCas9-pB is still a worthwhile system to research, because it offers target flexibility with computationally directed gRNA selection and cloning.

While Luo et al. began to form tools to allow isolation of targeted-only cells, they suggested that the next step would be to achieve specificity with no off-target integrations and greater efficiency [46].

## Highly Specific RNA-Guided Transposition in *E. coli*

A recent survey of bacterial and archaeal genomes revealed the presence of CRISPR-Cas genes within naturally occurring Tn7-like transposons [53]. The CRISPR-Cas genes encode the necessary components for CRISPR RNA processing and DNA binding but not for DNA cleavage. The transposons also contained CRISPR spacers, from which guide RNA could be made, containing target sequences from plasmids, phage, and bacterial chromosomal sequences. Remarkably, instances of transposon insertion next to these CRISPR-guided sequences were found. The authors suggest that the CRISPR-Cas components direct the insertion of the transposase near guide target sequences and facilitate horizontal transfer of the Tn7-like element. Because Tn7 and other Tn7-like transposon systems first require docking of the DBD before transposition can be activated, this system has the potential to overcome issues of off-target insertion that burden previous approaches. For targetable transposition to be strictly specific, the transposase must *require* the DBD for insertion.

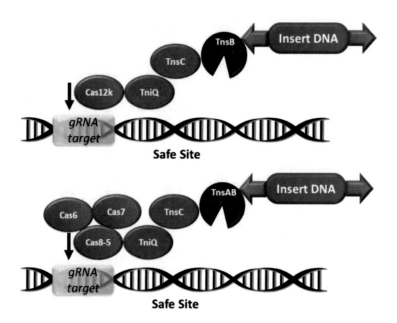

Figure 15. A. Cas12k binds to the target sequence guided by gRNA. A natural interaction between Cas12k and TniQ directs TniQ, TnsC, and the TnsB transposase to the target site and promote insertion. B. A multi-subunit Type I-F CRISPR–Cas Cascade system consisting of Cas6, Cas7, and Cas8-5 is guided by gRNA which also interacts with TniQ to direct the transposase to the RNA-guided target sequence.

Inspired by this finding, Strecker et al. characterized two Tn7-like transposons from Cyanobacteria *Scytonema hofmanni* and *Anabaena cylindrica* that each encoded a Type V-K CRISPR-Cas single-protein effector called Cas12k [47]. The authors termed the CRISPR-associated Transposases: CAST. To investigate the capability of these transposons to mediate RNA-guided transposition, *Cas12k* was co-expressed with the three transposase genes: *tnsB, tnsC, tniQ* in addition to a guide RNA and a transposon donor plasmid in *E. coli* (Figure 15A). Insertions were isolated from a target plasmid within a small window 60-66 bp and 49-56 bp downstream from the PAM for ShCAST and AcCAST, respectively. All insertions occurred unidirectionally and NGTN PAMs were preferred for both ShCAST and AcCAST systems. Similar to canonical Tn7, CAST transposon ends act in cis to prevent multiple insertions at the same site. Integration efficiency was reduced from ~60% to ~25% as the transposon size was increased from 0.5kb to 10kb, respectively. To exclude the requirement of bacterial host factors, in vitro targeting was demonstrated. The same purified protein components successfully mediated targeting in bacteria as well. The authors attempted to target 48 individual sites throughout the *E. coli* genome and successfully recovered insertions at 60% of those sites. Two sites in particular were targeted an impressive 80% of the time. A deep-sequencing analysis revealed that >50% of total insertions in the *E. coli* genome were on-target, with the remaining off-target sites enriched at actively transcribed genes.

A second study published in parallel demonstrated RNA-guided transposition of the Tn7-like transposon Tn*6677* from *Vibrio cholerae* [48]. This transposon encodes components from a Type I-F CRISPR-Cas system made up of three genes: *cas6*, *cas7*, and a natural *cas8-cas5* fusion. Similar to Cas12k, this complex contains DNA binding components but not cleavage components. The *tniQ* gene has been re-positioned to the CRISPR-Cas operon and the *tnsE* gene, which would usually direct insertion of the transposase, is absent. Co-expression of the four Tn7-like components *tnsA*, *tnsB*, *tnsC* and *tniQ* along with the three CRISPR-Cas genes and a guide RNA led to highly specific targeting (Figure 15B). Remarkably, >95% of integration events for the first guide RNA tested occurred 49-bp from the target site. Subsequent guide RNAs produced similar results with integrations occurring bidirectionally and deep sequencing revealing 95% targeting specificity for 16 guide RNAs. TniQ co-purified with Cas8, Cas7, and Cas6 and the authors proposed that TniQ links CRISPR- and transposon-associated machineries during DNA targeting and DNA integration. Increases in transposon size above 0.8 kb reduced integration efficiency and the CC dinucleotide PAM sequence was somewhat flexible. Taken together these two studies clearly demonstrate an exciting mechanism for highly specific RNA-guided transposition that someday might be used to insert DNA into the human genome.

## Pointing toward Modifications for "Site Required" Targetable Transposase for Human Applications

Despite successes in targeting specific endogenous locations, all targetable transposase approaches performed in mammalian cells to date have been hindered by frequent off-target insertion events. This is due to the fact that a fully functional transposase, despite being tethered to a DBD, retains the ability to bind to DNA at off-target sites and, therefore, does not rely on its associated DBD for integration. A likely scenario is that the transposase contacts numerous locations in the genome before binding to the target sequence and thus primarily performs off-target integrations.

A modified transposase could be envisioned in which mutations in the transposase's own DBD impair binding to DNA. A fusion of a custom DBD, such as a TALE, to this modified transposase could theoretically provide the only means for the transposase to bind to DNA and, by extension, allow integration near the binding site.

This hypothesis was tested by a "top down" approach using scanning alanine mutagenesis of conserved positively charged amino acids near the pB catalytic core [49]. By screening for mutants that retain excision activity of the transposon but are deficient for insertion of the transposon, three "excision-only" pB mutations were identified. It could be reasoned that a transposase that is no longer capable of integrating may have lost its ability to bind DNA. Fusion of custom zinc finger DBDs, designed to bind the ROSA26 and GULOP safe harbors, to the excision-only pB did not result in an integration-competent pB (Figure 15A). However, a different pB mutant which displayed partial integration of < 10% compared to unmodified pB, did recover full integration ability upon the fusion of these DBDs. Unfortunately, although integration activity was restored, these fusion transposases failed to target (Figure 15B). Nevertheless, these types of experiments will be crucial for the future development of this technology. Clearly, additional modifications, perhaps unrelated to the mutations identified in the excision-only pB, will need to be discovered in order to generate a truly "site required" targetable transposase.

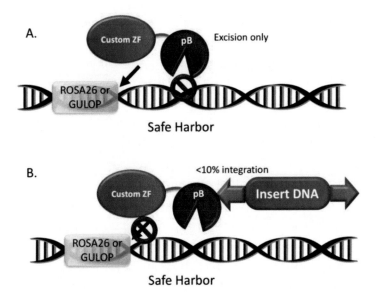

Figure 16. Fusion of custom zinc finger DBDs designed to bind ROSA26 or GULOP to excision only pB (A) and a partial excision (< 10% integration abilities) pB (B) did not result in integration or targeted integration.

## The First RNA-Guided Transposition in Human Cells

Hew et al. fused hyPBase to the highly-specific, catalytically-dead SpCas9-HF1 (dCas9) with a long flexible linker [50]. Inactivating mutations were introduced that disrupt the cutting but not the binding of dCas9. gRNAs were designed to target the *CCR5* safe harbor, which is a clinically-relevant site as described above (Figure 16). Four different gRNAs were expressed from the helper plasmid backbone to assist in increasing on-target efficiency.

A major challenge to pB transposition is off-target integration. To attempt to combat this, Hew et al. introduced mutations to the native DBD of pB to produce three variants with the goal of reducing nonspecific binding and increasing binding preference to dCas9. The PB variant had the seven hyPBase hyperactive mutations described above but no mutations in the pB DBD. Variants H2 and H3 have two and three mutations in the native DBD, respectively, with mutations in residues believed to play a role in DNA binding.

Hew et al. first tested integration with no fusion and found reduced integration levels for H2 and H3, which is reasonable considering their expected impaired ability to bind DNA. Then, they tested integration with fusion constructs. They found 10% integration for pB, 8% for H2, and 4% for H3. Also, the gRNAs targeted their intended sequences with 14-43% efficiency.

Finally, they transfected helper transposase expression plasmids with 4-guide or 8-guide combinations with transposon donor plasmid. They used neomycin selection to recover insertions and nested genomic PCR to recover targeted insertions. They used eight unique gRNAs tiled across the first intron of *CCR5* to direct insertions of dCas9-pB fusion constructs. They engineered a cloning strategy to simultaneously add multiple gRNAs in one step and found that increasing the number of gRNAs from four to eight improved targeting efficiency.

They recovered 17 independent insertions sites. The greatest number of insertions was for the 8-guide H2 and H3 variants that contained mutations in the native DBD, with over half from dCas9-H2-8guide and 59% at a single TTAA hotspot. This was probably due to reduced off-target binding of H2. Multiple guides likely contributed to a greater concentration of transposase proteins near the target sequence and increased the likelihood of hitting the insertion site. The clustering of individual pB monomers could support dimerization and facilitate integration at the target.

The dCas9-H2-8guide was used to generate single targeted clones. A two-step pooling process was used to identify positive pools of cells that were single-cell sorted. Genomic PCR was used to recover two positive clonal lines with a single insertion and stably expressed transgene. The efficiency of targeting was estimated to be 0.06% of cells.

The low specificity of targeting represents the greatest challenge but Hew et al.'s research represents the foundation for future experiments. Ultimately, this is the first evidence that RNA can be used to direct transposon insertions in human cells. While there was a low specificity,

Hew et al. effectively introduced a new tool that can be used to deliver large inserts away from undesirable sequences. Future studies could examine if alternate pB mutation also support RNA-guided transposition [50].

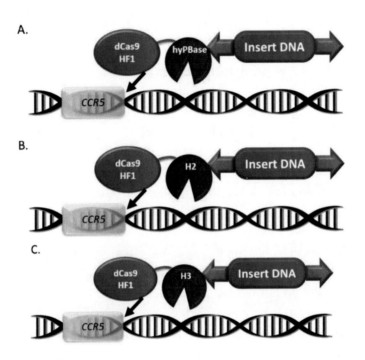

Figure 17. A. Fusion of hyperactive hypBase to dCas9 and expression of either 4 or 8 gRNAs resulted in on-target insertions to the CCR5 gene in human cells. PiggyBac variants with reduced DNA binding resulting from two mutations, H2 (B) or three mutations, H3 (C) improved RNA-guided targeting over non-mutated pB.

## FUTURE DIRECTIONS

Critical barriers exist for all current gene addition technologies. For example, untargeted virus-based approaches can disrupt important genes and are limited by cargo size and immune response [32, 54-62]. While major hurdles need to be "leapt over" before targeting transposition with jumping genes becomes clinically available, the new stimulation of research to reduce off-target integration will help to achieve greater specificity of the vectors. Ideally, gene transfer would occur exclusively at genomic "safe harbors" and nowhere else. Generating an efficient targetable vector will help to avoid oncogene activation, tumor suppressor inactivation, or silencing following integration into non-permissible genomic regions.

While these studies have taken crucial steps in the development of this technology, there is great room for optimization of current transposon targeting approaches and various options for future studies. First, defining the ideal spacing between the DBD and intended insertion site could improve efficiency. Additionally, selection of a highly specific DBD that binds a single site is imperative. Titrating the amount of transposase could reduce off-target insertions because high concentrations of a chimeric transposase would lead to the DBD recognition sites becoming continually occupied. This may cause the additional unbound proteins to stray to distant genomic locations.

In addition, a mechanism that tied integration to specific DNA binding could ideally be developed. For example, TALEN pairs act as a dimer which generally do not interact in solution. The FokI nuclease domain is only active when two components, each consisting of FokI and the TALE DBD, bind in close proximity at the target DNA [63]. Only then does FokI dimerize and cause a double strand break.

A similar model could be devised for a transposase-DBD fusion in which the dimerization domain is first inactivated by intentional mutations. Like TALENs, the transposase would not be able to dimerize and would not be active at off-target sites, however, binding of two associated DBDs in close proximity could theoretically join the transposase dimers and enable integration. An alternative model, mentioned above, could be to inactivate the transposase's own DBD so that it would become dependent on the introduced chimeric DBD for localization to DNA [49]. These "top-down" rational approaches are tempting, however, an unpredicted mechanism may arise from "bottom up" random mutational experiments.

Future studies that analyze potential mutations to create a transposase unable to autonomously bind in addition to using RNA to guide the transposase to a specific target would help overcome the greatest challenge to this technology. Possible strategies include screening for random or rationally-designed mutations that encourage co-dependence of a binding protein, such as dCas9 and pB for insertion. In addition, using hyPBase or a novel transposase with increased efficiency could improve a targeted transposon system. Ultimately, additional studies are warranted so that a truly site-specific transposase can be developed for improved gene editing and therapies in the future.

## CONCLUSION

As our understanding of genetic disease becomes more clear, the tools that give us the ability to correct genetic deficiencies become increasingly valuable. A vector that only targets a specific sequence at high efficiency would be a highly desirable clinical tool. This approach could contribute to modification of cell lines, gene therapy, and transgenic animal generation. The studies covered in this review will be crucial for this important development. So far, the research shows that targeted insertions occur more frequently with optimal target spacing, the use of an optimal linker, transient interaction, and specific fused transposases [36, 39, 41, 42]. In addition, an RNA-guided transposase holds potential for producing specific on-target insertions in human cells that are easily programmed [50].

By improving on the safety of these innovative methods, new applications for gene therapy will become possible. Dr. McClintock's glimpse of transposing genetic elements that shocked the scientific community instigated the development of these tools that may someday enable gene therapy to become a universal practice for combating, not only terminal illnesses, but a broad-spectrum of disease.

**REFERENCES**

[1]    Ravindran S. Barbara McClintock and the discovery of jumping genes. *Proc Natl Acad Sci U S A.* 2012;109(50):20198-9. Epub 2012/12/14. doi: 10.1073/pnas.1219372109. PubMed PMID: 23236127; PubMed Central PMCID: PMCPMC3528533.

[2]    Hudecek M, Izsvak Z, Johnen S, Renner M, Thumann G, Ivics Z. Going non-viral: the Sleeping Beauty transposon system breaks on through to the clinical side. *Crit Rev Biochem Mol Biol.* 2017;52(4): 355-80. Epub 2017/04/13. doi: 10.1080/10409238.2017.1304354. PubMed PMID: 28402189.

[3]    Munoz-Lopez M, Garcia-Perez JL. DNA transposons: nature and applications in genomics. *Curr Genomics.* 2010;11(2):115-28. Epub 2010/10/05. doi: 10.2174/138920210790886871. PubMed PMID: 20885819; PubMed Central PMCID: PMCPMC2874221.

[4]    Woodard LE, Wilson MH. piggyBac-ing models and new therapeutic strategies. *Trends Biotechnol.* 2015;33(9):525-33. Epub 2015/07/28. doi: 10.1016/j.tibtech.2015.06.009. PubMed PMID: 26211958; PubMed Central PMCID: PMCPMC4663986.

[5]    Yusa K. piggyBac Transposon. *Microbiol Spectr.* 2015;3(2): MDNA3-0028-2014. Epub 2015/06/25. doi: 10.1128/microbiolspec. MDNA3-0028-2014. PubMed PMID: 26104701.

[6]    Ding S, Wu X, Li G, Han M, Zhuang Y, Xu T. Efficient transposition of the piggyBac (PB) transposon in mammalian cells and mice. *Cell.* 2005;122(3):473-83. doi: 10.1016/j.cell.2005.07.013. PubMed PMID: 16096065.

[7]    Mates L, Chuah MK, Belay E, Jerchow B, Manoj N, Acosta-Sanchez A, Grzela DP, Schmitt A, Becker K, Matrai J, Ma L, Samara-Kuko E, Gysemans C, Pryputniewicz D, Miskey C, Fletcher B, Vandendriessche T, Ivics Z, Izsvak Z. Molecular evolution of a novel hyperactive Sleeping Beauty transposase enables robust stable gene transfer in vertebrates. *Nat Genet.* 2009;41(6):753-61. Epub 2009/05/05. doi: ng.343 [pii] 10.1038/ng.343. PubMed PMID: 19412179.

[8] Yusa K, Zhou L, Li MA, Bradley A, Craig NL. A hyperactive piggyBac transposase for mammalian applications. *Proc Natl Acad Sci U S A*. 2011;108(4):1531-6. Epub 2011/01/06. doi: 10.1073/pnas. 1008322108. PubMed PMID: 21205896; PubMed Central PMCID: PMCPMC3029773.

[9] Doherty JE, Huye LE, Yusa K, Zhou L, Craig NL, Wilson MH. Hyperactive piggyBac gene transfer in human cells and *in vivo*. *Hum Gene Ther*. 2012;23(3):311-20. Epub 2011/10/14. doi: 10.1089/hum. 2011.138. PubMed PMID: 21992617; PubMed Central PMCID: PMCPMC3300075.

[10] Rad R, Rad L, Wang W, Cadinanos J, Vassiliou G, Rice S, Campos LS, Yusa K, Banerjee R, Li MA, de la Rosa J, Strong A, Lu D, Ellis P, Conte N, Yang FT, Liu P, Bradley A. PiggyBac transposon mutagenesis: a tool for cancer gene discovery in mice. Science. 2010;330(6007):1104-7. Epub 2010/10/16. doi: 10.1126/science. 1193004. PubMed PMID: 20947725.

[11] Ikadai H, Shaw Saliba K, Kanzok SM, McLean KJ, Tanaka TQ, Cao J, Williamson KC, Jacobs-Lorena M. Transposon mutagenesis identifies genes essential for Plasmodium falciparum gametocytogenesis. *Proc Natl Acad Sci U S A*. 2013;110(18):E1676-84. Epub 2013/04/11. doi: 10.1073/pnas.1217712110. PubMed PMID: 23572579; PubMed Central PMCID: PMCPMC3645567.

[12] Yant SR, Meuse L, Chiu W, Ivics Z, Izsvak Z, Kay MA. Somatic integration and long-term transgene expression in normal and haemophilic mice using a DNA transposon system. *Nat Genet*. 2000;25(1):35-41. Epub 2000/05/10. doi: 10.1038/75568. PubMed PMID: 10802653.

[13] Saridey SK, Liu L, Doherty JE, Kaja A, Galvan DL, Fletcher BS, Wilson MH. PiggyBac transposon-based inducible gene expression in vivo after somatic cell gene transfer. Molecular therapy : *The journal of the American Society of Gene Therapy*. 2009;17(12):2115-20. Epub 2009/10/08. doi: 10.1038/mt.2009.234. PubMed PMID: 19809403; PubMed Central PMCID: PMC2814386.

[14]   Yusa K, Rad R, Takeda J, Bradley A. Generation of transgene-free induced pluripotent mouse stem cells by the piggyBac transposon. *Nat Methods.* 2009;6(5):363-9. Epub 2009/04/02. doi: 10.1038/nmeth.1323. PubMed PMID: 19337237; PubMed Central PMCID: PMCPMC2677165.

[15]   Owens JB, Urschitz J, Stoytchev I, Dang NC, Stoytcheva Z, Belcaid M, Maragathavally KJ, Coates CJ, Segal DJ, Moisyadi S. Chimeric piggyBac transposases for genomic targeting in human cells. *Nucleic Acids Res.* 2012;40(14):6978-91. Epub 2012/04/12. doi: 10.1093/nar/gks309. PubMed PMID: 22492708; PubMed Central PMCID: PMC3413120.

[16]   Wu SC, Meir YJ, Coates CJ, Handler AM, Pelczar P, Moisyadi S, Kaminski JM. piggyBac is a flexible and highly active transposon as compared to Sleeping Beauty, Tol2, and Mos1 in mammalian cells. *Proc Natl Acad Sci U S A.* 2006;103(41):15008-13. PubMed PMID: 17005721.

[17]   Marh J, Stoytcheva Z, Urschitz J, Sugawara A, Yamashiro H, Owens JB, Stoytchev I, Pelczar P, Yanagimachi R, Moisyadi S. Hyperactive self-inactivating piggyBac for transposase-enhanced pronuclear microinjection transgenesis. *Proceedings of the National Academy of Sciences of the United States of America.* 2012;109(47):19184-9. doi: 10.1073/pnas.1216473109. PubMed PMID: 23093669; PubMed Central PMCID: PMC3511126.

[18]   Tipanee J, VandenDriessche T, Chuah MK. Transposons: Moving Forward from Preclinical Studies to Clinical Trials. *Hum Gene Ther.* 2017;28(11):1087-104. Epub 2017/09/19. doi: 10.1089/hum. 2017.128. PubMed PMID: 28920716.

[19]   Manuri PV, Wilson MH, Maiti SN, Mi T, Singh H, Olivares S, Dawson MJ, Huls H, Lee DA, Rao PH, Kaminski JM, Nakazawa Y, Gottschalk S, Kebriaei P, Shpall EJ, Champlin RE, Cooper LJ. piggyBac transposon/transposase system to generate CD19-specific T cells for the treatment of B-lineage malignancies. *Hum Gene Ther.* 2010;21(4):427-37. Epub 2009/11/13. doi: 10.1089/hum.2009.114.

PubMed PMID: 19905893; PubMed Central PMCID: PMCPMC2938363.

[20] He J, Zhang Z, Lv S, Liu X, Cui L, Jiang D, Zhang Q, Li L, Qin W, Jin H, Qian Q. Engineered CAR T cells targeting mesothelin by piggyBac transposon system for the treatment of pancreatic cancer. *Cell Immunol.* 2018;329:31-40. Epub 2018/06/04. doi: 10.1016/j. cellimm.2018.04.007. PubMed PMID: 29859625.

[21] Bishop DC, Xu N, Tse B, O'Brien TA, Gottlieb DJ, Dolnikov A, Micklethwaite KP. PiggyBac-Engineered T Cells Expressing CD19-Specific CARs that Lack IgG1 Fc Spacers Have Potent Activity against B-ALL Xenografts. *Mol Ther.* 2018;26(8):1883-95. Epub 2018/06/05. doi: 10.1016/j.ymthe.2018.05.007. PubMed PMID: 29861327; PubMed Central PMCID: PMCPMC6094355.

[22] Li MA, Turner DJ, Ning Z, Yusa K, Liang Q, Eckert S, Rad L, Fitzgerald TW, Craig NL, Bradley A. Mobilization of giant piggyBac transposons in the mouse genome. *Nucleic acids research.* 2011. Epub 2011/09/29. doi: 10.1093/nar/gkr764. PubMed PMID: 21948799.

[23] Marwick C. FDA halts gene therapy trials after leukaemia case in France. *BMJ.* 2003;326(7382):181. PubMed PMID: 12543825; PubMed Central PMCID: PMC1125057.

[24] Suzuki T, Shen H, Akagi K, Morse HC, Malley JD, Naiman DQ, Jenkins NA, Copeland NG. New genes involved in cancer identified by retroviral tagging. *Nat Genet.* 2002;32(1):166-74. doi: 10.1038/ng949. PubMed PMID: 12185365.

[25] Deichmann A, Hacein-Bey-Abina S, Schmidt M, Garrigue A, Brugman MH, Hu J, Glimm H, Gyapay G, Prum B, Fraser CC, Fischer N, Schwarzwaelder K, Siegler ML, de Ridder D, Pike-Overzet K, Howe SJ, Thrasher AJ, Wagemaker G, Abel U, Staal FJ, Delabesse E, Villeval JL, Aronow B, Hue C, Prinz C, Wissler M, Klanke C, Weissenbach J, Alexander I, Fischer A, von Kalle C, Cavazzana-Calvo M. Vector integration is nonrandom and clustered and influences the fate of lymphopoiesis in SCID-X1 gene therapy. *The Journal of clinical investigation.* 2007;117(8):2225-32. Epub

2007/08/03. doi: 10.1172/JCI31659. PubMed PMID: 17671652; PubMed Central PMCID: PMC1934585.

[26]  Cox DB, Platt RJ, Zhang F. Therapeutic genome editing: prospects and challenges. *Nature medicine.* 2015;21(2):121-31. doi: 10.1038/nm.3793. PubMed PMID: 25654603; PubMed Central PMCID: PMC4492683.

[27]  Lieber MR. The mechanism of double-strand DNA break repair by the nonhomologous DNA end-joining pathway. *Annual review of biochemistry.* 2010;79:181-211. doi: 10.1146/annurev.biochem. 052308.093131. PubMed PMID: 20192759; PubMed Central PMCID: PMC3079308.

[28]  Fung H, Weinstock DM. Repair at single targeted DNA double-strand breaks in pluripotent and differentiated human cells. *PloS one.* 2011;6(5):e20514. doi: 10.1371/journal.pone.0020514. PubMed PMID: 21633706; PubMed Central PMCID: PMC3102116.

[29]  Orthwein A, Noordermeer SM, Wilson MD, Landry S, Enchev RI, Sherker A, Munro M, Pinder J, Salsman J, Dellaire G, Xia B, Peter M, Durocher D. A mechanism for the suppression of homologous recombination in G1 cells. *Nature.* 2015;528(7582):422-6. doi: 10.1038/nature16142. PubMed PMID: 26649820; PubMed Central PMCID: PMC4880051.

[30]  Kosicki M, Tomberg K, Bradley A. Repair of double-strand breaks induced by CRISPR-Cas9 leads to large deletions and complex rearrangements. *Nature biotechnology.* 2018;36(8):765-71. doi: 10.1038/nbt.4192. PubMed PMID: 30010673.

[31]  Anderson CD, Urschitz J, Khemmani M, Owens JB, Moisyadi S, Shohet RV, Walton CB. Ultrasound directs a transposase system for durable hepatic gene delivery in mice. *Ultrasound Med Biol.* 2013;39(12):2351-61. doi: 10.1016/j.ultrasmedbio.2013.07.002. PubMed PMID: 24035623; PubMed Central PMCID: PMC3838570.

[32]  Li MA, Turner DJ, Ning Z, Yusa K, Liang Q, Eckert S, Rad L, Fitzgerald TW, Craig NL, Bradley A. Mobilization of giant piggyBac transposons in the mouse genome. *Nucleic acids research.*

2011;39(22):e148. Epub 2011/09/29. doi: 10.1093/nar/gkr764. PubMed PMID: 21948799; PubMed Central PMCID: PMC3239208.

[33]  Kaminski JM, Huber MR, Summers JB, Ward MB. Design of a nonviral vector for site-selective, efficient integration into the human genome. *The FASEB journal : Official publication of the Federation of American Societies for Experimental Biology.* 2002;16(10):1242-7. Epub 2002/08/03. doi: 10.1096/fj.02-0127hyp. PubMed PMID: 12153992.

[34]  Coates CJ, Kaminski JM, Summers JB, Segal DJ, Miller AD, Kolb AF. Site-directed genome modification: derivatives of DNA-modifying enzymes as targeting tools. *Trends Biotechnol.* 2005;23(8):407-19. Epub 2005/07/05. doi: 10.1016/j.tibtech.2005. 06.009. PubMed PMID: 15993503.

[35]  Peters JE, Craig NL. Tn7: smarter than we thought. *Nat Rev Mol Cell Biol.* 2001;2(11):806-14. doi: 10.1038/35099006. PubMed PMID: 11715047.

[36]  Feng X, Bednarz AL, Colloms SD. Precise targeted integration by a chimaeric transposase zinc-finger fusion protein. *Nucleic acids research.* 2010;38(4):1204-16. Epub 2009/12/08. doi: 10.1093/nar/ gkp1068. PubMed PMID: 19965773; PubMed Central PMCID: PMC2831304.

[37]  Szabo M, Muller F, Kiss J, Balduf C, Strahle U, Olasz F. Transposition and targeting of the prokaryotic mobile element IS30 in zebrafish. *FEBS Lett.* 2003;550(1-3):46-50. Epub 2003/08/26. PubMed PMID: 12935884.

[38]  Maragathavally KJ, Kaminski JM, Coates CJ. Chimeric Mos1 and piggyBac transposases result in site-directed integration. *Faseb J.* 2006. PubMed PMID: 16877528.

[39]  Yant SR, Huang Y, Akache B, Kay MA. Site-directed transposon integration in human cells. *Nucleic Acids Res.* 2007;35(7):e50. Epub 2007/03/09. doi: 10.1093/nar/gkm089. PubMed PMID: 17344320; PubMed Central PMCID: PMC1874657.

[40]  Ivics Z, Katzer A, Stuwe EE, Fiedler D, Knespel S, Izsvak Z. Targeted Sleeping Beauty transposition in human cells. Molecular

therapy : *The journal of the American Society of Gene Therapy.*
2007;15(6):1137-44. Epub 2007/04/12. doi: 10.1038/sj.mt.6300169.
PubMed PMID: 17426709.

[41]   Kettlun C, Galvan DL, George AL, Jr., Kaja A, Wilson MH.
       Manipulating piggyBac Transposon Chromosomal Integration Site
       Selection in Human Cells. *Mol Ther.* 2011;19(9):1636-44. Epub
       2011/07/07. doi: 10.1038/mt.2011.129. PubMed PMID: 21730970.

[42]   Voigt K, Gogol-Doring A, Miskey C, Chen W, Cathomen T, Izsvak
       Z, Ivics Z. Retargeting sleeping beauty transposon insertions by
       engineered  zinc  finger  DNA-binding  domains.  *Mol  Ther.*
       2012;20(10):1852-62.  doi:  10.1038/mt.2012.126.  PubMed  PMID:
       22776959; PubMed Central PMCID: PMC3464645.

[43]   Ammar I, Gogol-Doring A, Miskey C, Chen W, Cathomen T, Izsvak
       Z, Ivics Z. Retargeting transposon insertions by the adeno-associated
       virus Rep protein. *Nucleic Acids Res.* 2012;40(14):6693-712. Epub
       2012/04/24. doi: 10.1093/nar/gks317. PubMed PMID: 22523082;
       PubMed Central PMCID: PMC3413126.

[44]   Wang H, Mayhew D, Chen X, Johnston M, Mitra RD. "Calling
       cards" for DNA-binding proteins in mammalian cells. *Genetics.*
       2012;190(3):941-9.   doi:   10.1534/genetics.111.137315.   PubMed
       PMID: 22214611; PubMed Central PMCID: PMC3296256.

[45]   Owens JB, Mauro D, Stoytchev I, Bhakta MS, Kim MS, Segal DJ,
       Moisyadi S. Transcription activator like effector (TALE)-directed
       piggyBac transposition in human cells. *Nucleic Acids Res.*
       2013;41(19):9197-207. Epub 2013/08/08. doi: 10.1093/nar/gkt677.
       PubMed PMID: 23921635; PubMed Central PMCID: PMC3799441.

[46]   Luo W, Galvan DL, Woodard LE, Dorset D, Levy S, Wilson MH.
       Comparative analysis of chimeric ZFP-, TALE- and Cas9-piggyBac
       transposases for integration into a single locus in human cells.
       *Nucleic acids research.* 2017;45(14):8411-22. doi: 10.1093/nar/
       gkx572. PubMed PMID: 28666380; PubMed Central PMCID:
       PMC5737283.

[47]   Strecker J, Ladha A, Gardner Z, Schmid-Burgk JL, Makarova KS,
       Koonin EV, Zhang F. RNA-guided DNA insertion with CRISPR-

associated transposases. *Science.* 2019. doi: 10.1126/science.aax 9181. PubMed PMID: 31171706.

[48] Klompe SE, Vo PLH, Halpin-Healy TS, Sternberg SH. Transposon-encoded CRISPR-Cas systems direct RNA-guided DNA integration. *Nature.* 2019. doi: 10.1038/s41586-019-1323-z. PubMed PMID: 31189177.

[49] Li X, Burnight ER, Cooney AL, Malani N, Brady T, Sander JD, Staber J, Wheelan SJ, Joung JK, McCray PB, Jr., Bushman FD, Sinn PL, Craig NL. piggyBac transposase tools for genome engineering. *Proceedings of the National Academy of Sciences of the United States of America.* 2013. Epub 2013/06/01. doi: 10.1073/pnas. 1305987110. PubMed PMID: 23723351.

[50] Hew BE, Sato R, Mauro D, Stoytchev I, Owens JB. RNA-guided *piggyBac* transposition in human cells. *Synth Biol (Oxf).* 2019;4(1):ysz018. Epub 2019/07/02 doi: 10.1093/synbio/ysz018. PubMed PMID: 31355344

[51] Ivics Z, Hackett PB, Plasterk RH, Izsvak Z. Molecular reconstruction of Sleeping Beauty, a Tc1-like transposon from fish, and its transposition in human cells. *Cell.* 1997;91(4):501-10. PubMed PMID: 9390559.

[52] Maier DA, Brennan AL, Jiang S, Binder-Scholl GK, Lee G, Plesa G, Zheng Z, Cotte J, Carpenito C, Wood T, Spratt SK, Ando D, Gregory P, Holmes MC, Perez EE, Riley JL, Carroll RG, June CH, Levine BL. Efficient Clinical Scale Gene Modification via Zinc Finger Nuclease-Targeted Disruption of the HIV Co-receptor CCR5. *Human gene therapy.* 2013;24(3):245-58. Epub 2013/01/31. doi: 10.1089/hum.2012.172. PubMed PMID: 23360514.

[53] Peters JE, Makarova KS, Shmakov S, Koonin EV. Recruitment of CRISPR-Cas systems by Tn7-like transposons. *Proceedings of the National Academy of Sciences of the United States of America.* 2017;114(35):E7358-E66. doi: 10.1073/pnas.1709035114. PubMed PMID: 28811374; PubMed Central PMCID: PMC5584455.

[54] Daniel R, Smith JA. Integration site selection by retroviral vectors: molecular mechanism and clinical consequences. *Human gene*

*therapy*. 2008;19(6):557-68. Epub 2008/06/07. doi: 10.1089/hum. 2007.148. PubMed PMID: 18533894; PubMed Central PMCID: PMC2940482.

[55] Mitchell RS, Beitzel BF, Schroder AR, Shinn P, Chen H, Berry CC, Ecker JR, Bushman FD. Retroviral DNA integration: ASLV, HIV, and MLV show distinct target site preferences. *PLoS biology*. 2004;2(8):E234. Epub 2004/08/18. doi: 10.1371/journal.pbio. 0020234. PubMed PMID: 15314653; PubMed Central PMCID: PMC509299.

[56] Hacein-Bey-Abina S, Garrigue A, Wang GP, Soulier J, Lim A, Morillon E, Clappier E, Caccavelli L, Delabesse E, Beldjord K, Asnafi V, MacIntyre E, Dal Cortivo L, Radford I, Brousse N, Sigaux F, Moshous D, Hauer J, Borkhardt A, Belohradsky BH, Wintergerst U, Velez MC, Leiva L, Sorensen R, Wulffraat N, Blanche S, Bushman FD, Fischer A, Cavazzana-Calvo M. Insertional oncogenesis in 4 patients after retrovirus-mediated gene therapy of SCID-X1. *J Clin Invest*. 2008;118(9):3132-42. Epub 2008/08/09. doi: 10.1172/JCI35700. PubMed PMID: 18688285; PubMed Central PMCID: PMC2496963.

[57] Howe SJ, Mansour MR, Schwarzwaelder K, Bartholomae C, Hubank M, Kempski H, Brugman MH, Pike-Overzet K, Chatters SJ, de Ridder D, Gilmour KC, Adams S, Thornhill SI, Parsley KL, Staal FJ, Gale RE, Linch DC, Bayford J, Brown L, Quaye M, Kinnon C, Ancliff P, Webb DK, Schmidt M, von Kalle C, Gaspar HB, Thrasher AJ. Insertional mutagenesis combined with acquired somatic mutations causes leukemogenesis following gene therapy of SCID-X1 patients. *J Clin Invest*. 2008;118(9):3143-50. Epub 2008/08/09. doi: 10.1172/JCI35798. PubMed PMID: 18688286; PubMed Central PMCID: PMC2496964.

[58] Hareendran S, Balakrishnan B, Sen D, Kumar S, Srivastava A, Jayandharan GR. Adeno-associated virus (AAV) vectors in gene therapy: immune challenges and strategies to circumvent them. *Rev Med Virol*. 2013;23(6):399-413. doi: 10.1002/rmv.1762. PubMed PMID: 24023004.

[59] Basner-Tschakarjan E, Mingozzi F. Cell-Mediated Immunity to AAV Vectors, Evolving Concepts and Potential Solutions. *Front Immunol.* 2014;5:350. doi: 10.3389/fimmu.2014.00350. PubMed PMID: 25101090; PubMed Central PMCID: PMC4107954.

[60] Rogers GL, Martino AT, Aslanidi GV, Jayandharan GR, Srivastava A, Herzog RW. Innate Immune Responses to AAV Vectors. *Front Microbiol.* 2011;2:194. doi: 10.3389/fmicb.2011.00194. PubMed PMID: 21954398; PubMed Central PMCID: PMC3175613.

[61] Boutin S, Monteilhet V, Veron P, Leborgne C, Benveniste O, Montus MF, Masurier C. Prevalence of serum IgG and neutralizing factors against adeno-associated virus (AAV) types 1, 2, 5, 6, 8, and 9 in the healthy population: implications for gene therapy using AAV vectors. *Human gene therapy.* 2010;21(6):704-12. doi: 10.1089/hum.2009.182. PubMed PMID: 20095819.

[62] Mingozzi F, High KA. Therapeutic in vivo gene transfer for genetic disease using AAV: progress and challenges. *Nat Rev Genet.* 2011;12(5):341-55. doi: 10.1038/nrg2988. PubMed PMID: 21499295.

[63] Szczepek M, Brondani V, Buchel J, Serrano L, Segal DJ, Cathomen T. Structure-based redesign of the dimerization interface reduces the toxicity of zinc-finger nucleases. *Nature biotechnology.* 2007;25(7):786-93. Epub 2007/07/03. doi: 10.1038/nbt1317. PubMed PMID: 17603476.

In: Gene Delivery  ISBN: 978-1-53616-268-4
Editor: Vanessa Zimmer  © 2019 Nova Science Publishers, Inc.

*Chapter 4*

# EFFICIENT DNA TRANSFECTION IN PROTISTS MEDIATED BY CELL-PENETRATING PEPTIDE

## *Betty Revon Liu[1],\*, Yue-Wern Huang[2] and Han-Jung Lee[3]*

[1]Department of Laboratory Medicine and Biotechnology,
Tzu Chi University, Hualien, Taiwan
[2]Department of Biological Sciences,
Missouri University of Science and Technology, Rolla, MO, US
[3]Department of Natural Resources and Environmental Studies,
National Dong Hwa University, Hualien, Taiwan

## ABSTRACT

Developing a useful and efficient DNA transfection method is always a concerning issue and necessary for study of specific molecules and their functions in individuals. However, most transfection methods utilized today are applied in mammalian cells such as embryonic cells

---

\* Corresponding Author's Email: brliu7447@gms.tcu.edu.tw.

and cell lines. Rare transfection studies are found in individual microscopic organisms such as paramecium and rotifer which belong to single-celled and multi-celled individuals, respectively. Here, we introduced cell-penetrating peptides (CPPs) as an efficient tool for DNA transfection. HR9, a designed CPP, containing nona-arginine flanked by cysteine and penta-histidine displayed a high penetrating ability in mammalian cells. Moreover, HR9 was able to internalize paramecia and rotifers which contain the pellicles and cuticles, respectively. DNAs were also delivered into these cells and organisms by HR9 and still contained the bioactivities. High viabilities of organisms and low cytotoxicities after HR9/DNA treatments illustrated that this CPP was harmless and could be a potent tool for DNA transfection.

## ABBREVIATIONS

| | |
|---|---|
| BFP | blue fluorescent protein |
| CPP | cell-penetrating peptide |
| CNPT | covalent and noncovalent protein transductions |
| DMSO | dimethyl sulfoxide |
| EGFP | enhanced green fluorescent protein |
| EtOH | alcohol |
| FITC | fluorescein isothiocyanate |
| GFP | green fluorescent protein |
| HIV-1 | human immunodeficiency virus type *1* |
| HR9 | nona-arginine flanked by cysteine and penta-histidine cell-penetrating peptide |
| MTT | 1-(4,5-dimethylthiazol-2-yl)-3,5-diphenylformazan |
| PBS | phosphate buffered saline |
| RFP | red fluorescent protein |
| SDs | standard deviations |
| Tat | transactivator of transcription |

# INTRODUCTION

A practical and efficient DNA transfection method is very important and required to study gene functions and regulations [1]. Multiple techniques of exogenous nucleic acid delivery have been developed, including viral and nonviral vector-mediated methods [2]. Retroviruses, lentiviruses, and adeno-associated viruses are three types of vectors which are commonly used in transient or stable transgene expression in mammalian cells [2]. Nonviral DNA delivery which is the prominent route can be further divided into two groups: chemical-based methods and physical techniques [2]. Calcium phosphate, cationic lipids, polymers, peptides, polysaccharides, and inorganic nanomaterials are utilized as the chemical-based methods, while physical techniques contain ultrasonic nebulization, microinjection, electroporation, and particle bombardment [2, 3]. Pros and cons of these gene delivery methods are being stated and the choice of a proper tool for transfected gene expression depends on the goal of a study and other factors [4]. Most DNA delivery systems were developed and applied in mammalian cells, including totipotent embryonic and pluripotent stem cells [4-6]. Few studies reported that these present transgenic tools were able to be applied in individual microscopic organisms. Therefore, development of safer, more efficient, easy-handled, and widely used methods for gene delivery are next steps for transgenic protists.

Cell-penetrating peptides (CPPs) are defined as short peptides which contained the ability to deliver cargoes to transport across plasma membrane [7]. The first reported CPP was Tat, which came from the transcription activator of the human immunodeficiency virus type *1* (HIV-*1*) [8]. The study of primary structure of the domain responsible for internalization in Tat revealed that the cationic residues, such as lysine and arginine, were the keys for the plasma membrane penetration [9]. Various CPPs were flourishingly developed later and able to be divided into three categories: cationic, amphipathic, and hydrophobic [9, 10]. CPPs were good shutters to take other biomolecules as cargoes into cells in covalent, noncovalent, or covalent and noncovalent protein transductions (CNPT) manners [11, 12]. The types of cargoes included proteins, DNAs, RNAs, small molecular drugs, and some inorganic particles [7, 9, 10, 12-16]. The sizes of cargoes were able to reach up to 200 nm in diameter [17]. Besides, the dosage of CPPs for bio-application was up to 100 μM and it still would not cause injury to cells [18]. Therefore, CPPs could be the good tools for therapeutic delivery into cells because of their high transduction efficiency, fast transduction rate, and low cytotoxicity.

Various DNA delivery methods were modified and applied to different organisms. Simple heat-shock gene transformations were commonly used in prokaryotes which had to be prepared as competent cells first [19]. Electroporation had to be modified to increase transfection efficiency in *Parabodo causatus* and *Euglena gracilis* [20, 21]. The viral RNA-based transfection method was reported in *Trichomonas vaginalis* [22]. However, no transfection methods could be used in all organisms. Besides, complicated procedures of competent cell preparation and gene-delivery system setting, low efficiency of successful transfection, and high cells/organisms injury become the drawbacks of these gene delivery systems. Our previous studies reported that CPPs possessed the ability to enter different organisms including prokaryotes, yeasts, insect cells, mammalian cells, plant tissues, paramecia, and rotifers [11, 15, 16, 23-31]. Here, we illustrated that one-CPP-one-protocol could be used for exogenous gene deliveries in mammalian cells, paramecia, and rotifers.

## MATERIALS AND METHODS

### Culture of Various Cells and Organisms

Human bronchoalveolar carcinoma A549 cells (American Type Culture Collection, Manassas, VA, USA; CCL-185) were grown in RPMI 1640 medium (Gibco, Invitrogen, Carlsbad, CA, USA) supplemented with 10% (v/v) bovine serum (Gibco). Cells were cultured in a humidified incubator with 5% $CO_2$ at 37°C as previously described [15].

Paramecia (*P. caudatum*) were grown in the culture medium of 1.25% (w/v) fresh lettuce juice [32] diluted with the Dryl's solution [33]. Young cells with 4–5 divisions were maintained at 23–25°C and seeded at a density of 50–100 paramecia per 100 µl in a well of 24-well plates for one day at room temperature before experiment.

Rotifers (*Brachionus calyciflorus*) (Bioprojects International Co.; Kaohsiung, Taiwan) were cultured in freshwater supplemented with the Fresh Chlorella V-12 (Bioprojects). The culture system was air-pumped at a rate of 0.1~0.3 L/min according to the manufacturer's instructions. Rotifers were seeded at a density of $1 \times 10^5$ in each well of 24-well plates and incubated in a shaker incubator at 25–28°C [16].

## Preparation of Plasmids and Peptides

The pEGFP-N1 plasmid (Clontech, Mountain View, CA, USA) containing the enhanced green fluorescent protein (*EGFP*) and the cytomegalovirus (CMV) promoter was used for CPP-mediated gene delivery and gene expression in transfected A549 cells. The pGFP-actin1-1 plasmid (kindly provided by Dr. Ilya N. Skovorodkin, University of Oulu, Finland) containing green fluorescent protein (GFP)-actin fusion gene is under the control of the *P. caudatum* α-tubulin promoter. The pCS2+ DsRed plasmid contains the coding regions of red fluorescent protein (*RFP*) (*DsRed1* with GenBank accession number JF330266) under the control of the simian cytomegalovirus immediate-early enhancer/promoter sequence (GenBank accession number U38308) [34, 35].

HR9 (CHHHHHRRRRRRRRRRHHHHHC) was synthesized as previously described [14]. The molecular mass of HR9 peptide is 3001.7 Daltons. HR9-FITC peptide containing the fluorescein isothiocyanate (FITC) at the N-terminus was chemically synthesized using solid-phase peptide synthesis and cross-linked with FITC by the FluoroTag FITC conjugation kit (Genomics, Taipei, Taiwan) as previously described [15]. HR9-FITC peptide contains a molecular mass of 3504 Daltons.

## Protein Transduction in Organisms

To monitor cellular uptake of HR9-FITC in mammalian cells, human A549 cells were treated with 10 µM of HR9-FITC for 1 h at 37°C followed by the treatment of Hoechst 33342 (Invitrogen) as previously described [14]. The cells were then washed with phosphate buffered saline (PBS) to remove superfluous peptides and dyes. The cells were monitored using a Nikon A1+ confocal fluorescent microscope with a magnification of 600× (Nikon Instruments Inc., Melville, NY, USA).

To observe cellular uptake of HR9-FITC in the protozoa, young paramecia were washed in water before treatment. Cells were treated with 7.15 µM of HR9-FITC for 1 hour at room temperature [25]. The cells were subsequently monitored right after incubation with 1% formaldehyde fixation using the BD Pathway 435 system (BD Biosciences, Franklin Lakes, NJ, USA). Another protozoa, rotifers, were treated with 6 µM of HR9-FITC for 1 h at 28°C as previously described [15, 25]. Live rotifers were washed and monitored without fixation. Fluorescent and bright-field images were recorded using a BD Pathway 435 System (BD Biosciences).

## CPP-Mediated Gene Delivery into Various Organisms

To observe gene delivery mediated by HR9 and the functional reporter gene assay, A549 cells were treated with either 3 µg of the pEGFP-N1 plasmid DNA only or the pEGFP-N1 plasmid DNA mixed with HR9 at the N/P ratio of 3. Cells treated with only medium served as a negative control. Cells were incubated for 10 min at 37°C. After that, solution was removed, and cells were washed with PBS thrice. A549 cells were then supplemented with 100 µl of 10% serum-containing medium and incubated at 37°C for 48 h. After 2 days, cells were stained with Hoechst 33342 and observed using a BD Pathway 435 system (BD Biosciences).

Three microgram of the circular pGFP-actin1-1 plasmid were mixed without or with HR9 at the N/P ratio of 3 in water at a final volume of 60 µl and incubated for 1 h at room temperature. The mock, pGFP-actin1-1 plasmid only, or HR9/pGFP-actin1-1 complexes were then dropped into cells and incubated for 30 min at room temperature. After the incubation, the cells were added with 1 ml of culture medium and incubated for additional 3 days at room temperature. Cells were monitored without fixation using the BD Pathway 435 system (BD Biosciences).

Rotifers were treated with water or 3 μg of the pCS2+ DsRed plasmid DNA alone as controls, while rotifers were treated with HR9/pCS2+ DsRed plasmid DNA complexes at the N/P ratio of 3 as an experimental group. After incubation for 10 min at 28°C, solution was removed, and rotifers were washed with freshwater thrice, followed by incubation at 28°C for 24 h. Rotifers were then observed using the Olympus BX51 inverted fluorescent microscope (Olympus, Center Valley, PA, USA).

The transfection efficiency of the HR9-mediated gene delivery and the gene expression intensity were determined as previously described [15, 16, 24, 25]. The relative intensities of fluorescent images from the functional gene assay were converted and quantified using the UN-SCAN-IT software (Silk Scientific, Orem, UT, USA).

## Confocal and Fluorescent Microscopy

Bright-field, GFP, and blue fluorescent protein (BFP) images were recorded using a BD Pathway 435 System (BD Biosciences) as previously described [14-16, 24, 25]. Excitation filters were set at 377/50, and 482/35 nm for blue and green, respectively. Emission filters were set at 435 LP (long-pass) and 536/40 nm for BFP and GFP channels, respectively. Transmitted light without the excitation filter, but with 536/40 nm emission filter, was used to observe cell morphology as bright-field images. The Olympus BX51 inverted fluorescent microscope (Olympus) was set with excitation at 531–554 nm and emission at 580-615 nm for RFP channel [16]. In mammalian cells observation, BFP, GFP, and bright-field images were detected using a Nikon A1+ confocal fluorescent microscope (Nikon Instruments Inc.) with excitation at 405 nm and emission at 450/50 nm for BFP and excitation at 488 nm and emission at 520–568 nm for GFP.

## Toxicity Measurement

The 1-(4,5-dimethylthiazol-2-yl)-3,5-diphenylformazan (MTT) assay was conducted to determine viability of A549 cells for 48 h, paramecia for 3 days, and rotifers for 24 h as previously described [15, 16, 25]. A549 and rotifers without any treatments served as the negative controls, while A549 cells and rotifers treated with 100% dimethyl sulfoxide (DMSO) served as the positive controls. In viability assay of paramecia, cells treated with culture medium alone served as the negative control, while paramecia treated with 70% alcohol (EtOH) served as a positive control. A549 cells, paramecia, and rotifers were treated with DNA only, HR9 only, and HR9/DNA complexes, respectively as the experimental groups

## Statistical Analysis

Data are presented as mean ± standard deviations (SDs). Statistical comparisons between the control and experimental groups were performed using the Student's $t$-test. Mean values and SDs were calculated for each groups examined in at least triplicate independent experiments. The level of statistical significance was set at $P < 0.05$ (*, †, α) or 0.01 (**, ††, αα).

## RESULTS

To assess whether HR9-FITC can enter mammalian cells, human A549 cells were treated with HR9-FITC peptide. No green signal was observed in the cells treated with PBS as a negative control (Figure 1A). However, strong green fluorescent signal was visualized in the cells treated with HR9-FITC at the GFP channel (Figure 1A). These results showed that this

CPP, HR9-FITC, possesses the cellular internalization activity in mammalian cells.

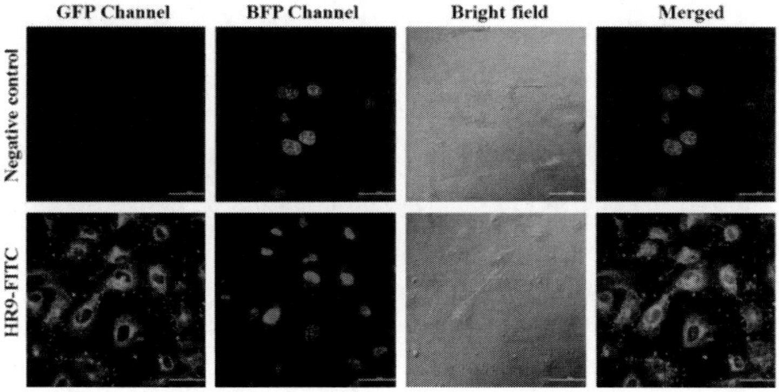

Figure 1. Protein transduction in various species. (A) Protein tranduction in human bronchoalveolar carcinoma A549 cells. Cells were treated with 10 µM HR9-FITC for 1 h. Cells without any treatments were served as the negative control. GFP, BFP channels and bright fields revealed the distributions of FITC-labeled HR9, nuclei, and cell morphologies. All images are obtained using a Nikon A1+ confocal fluorescent microscope with a magnification of 600×. (B) Protein transduction in the protozoa. Paramecia and rotifers were either treated with HR9-FITC or double-deionized water as the experimental groups and the negative control, respectively. GFP channels and bright fields illustrated the distributions of HR9-FITC and cell morphologies, and merged images were overlapped with GFP channels and bright fields. All images are obtained using a BD Pathway 435 confocal microscopic system with a magnification of 200×.

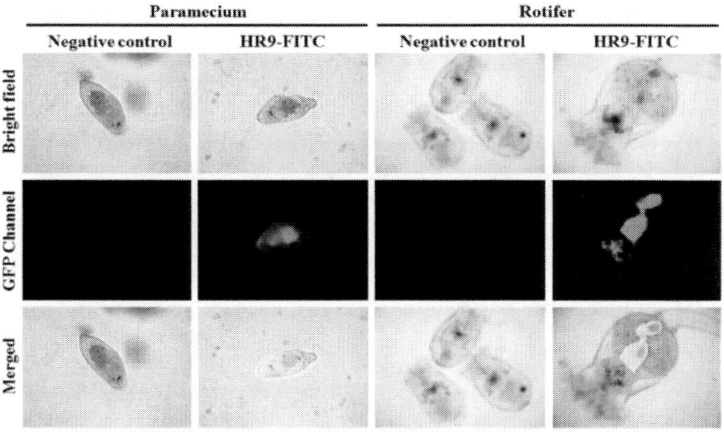

To demonstrate that CPPs can enter live organisms, HR9-FITC was used to incubate with paramecia and rotifers. We found that no signal was observed in the paramecia without any treatments as a control. On the other hand, green fluorescence was visualized in the paramecia treated with HR9-FITC at the GFP channel (Figure 1B). Similar results were shown in rotifers. No signal was detected in the rotifers treated with water as a control. On the contrary, green fluorescence was visualized in the rotifers treated with HR9-FITC (Figure 1B). These results demonstrated that HR9 can be internalized into both paramecia and rotifers (Table 1).

To determine plasmid DNA could be delivered by HR9 into organisms and the cargo DNA can be actively expressed after delivery, pEGFP-N1, pGFP-actin1-1, and pCS2+ DsRed were used for the functional gene assay in A549 cells, paramecia, and rotifers, respectively. Three microgram of plasmid DNA were mixed with HR9 at the N/P ratio of 3 and HR9/DNA complexes were added into these different organisms (Figure 2). After 48 h of incubation in A549 cells, no green fluorescent signal could be detected in the cells treated with PBS as a negative control and DNA only (Figure 2A). In contrast, green fluorescence was displayed in A549 cells treated with CPP/DNA complexes (Figure 2A), indicating that DNA transfection mediated by HR9 is efficient and gene expressed well in A549 cells (Figure 2B and 2C). In paramecia, no signals were observed while cells treated with either mock as a negative control or DNA only after 3 days recovery (Figure 2A). Bright fluorescence was easily observed in cells treated with HR9/DNA complexes (Figure 2A), revealing DNA was efficiently transported into paramecia and still contained its function (Figure 2B and 2C). In rotifers, extremely weak signals were detected in both the control group and the group treated with DNA only (Figure 2A). However, red fluorescence was visualized in the rotifers treated with HR9/DNA complexes (Figure 2A). These results demonstrate that HR9 is an effective transgenic carrier in Rotifera and transfected gene is activate (Figure 2B and 2C, Table 1).

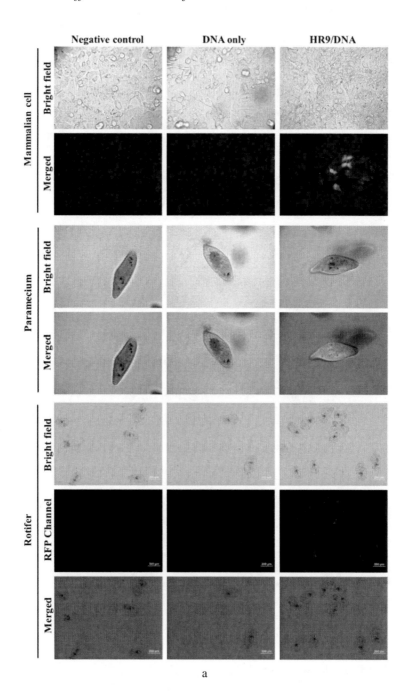

a

Figure 2. (Continued).

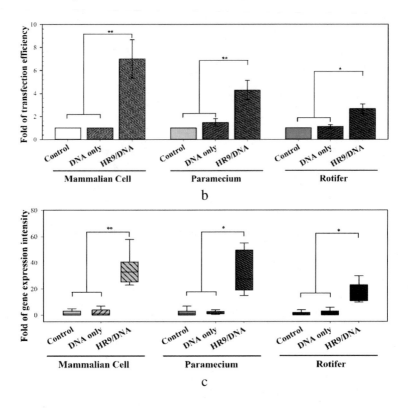

b

c

Figure 2. Plasmid DNA transfection via HR9 in different organisms. (A) Fluorescent microscopy of the HR9-mediated delivery of pEGFP-N1, pGFP-actin1-1, and pCS2+ DsRed into A549 cells, paramecia, and rotifers, respectively. The nuclei of A549 cells were stained with Hoechst 33342 and observed with BFP channel; cell-transfected gene expressions were observed with GFP channel in mammalian cells and paramecia. Merged images were obtained by the overlapping of GFP and BFP channels in the groups of mammalian cells as well as overlapping of GFP channel and bright field in the groups of paramecia, respectively. Transfected gene expression of pCS2+ DsRed was observed with RFP channel and merged images were obtained by RFP channel and bright field. Bright fields revealed the cell morphologies, shapes of paramecia, and appearances of rotifers. Images in mammalian cells and paramecia are procured using a BD Pathway 435 system at a magnification of 200×; while images in rotifers are shown using the Olympus BX51 inverted fluorescent microscope at a magnification of 200×. (B) Quantification of HR9-mediated gene transfection efficiency. DNA transfection efficiency was calculated from the fluorescent intensities at the GFP or RFP channels of figure 2A. (C) Quantification of gene expression intensity in HR9-mediated gene delivery. Gene expression intensities were converted from the digital image data of functional gene assay and analyzed by the UN-SCAN-IT software. Each experimental group was compared with the control group. Significant differences were determined at $P < 0.05$ (*) and $P < 0.01$ (**) between the control and various treatments in each organisms.

**Table 1. The comparison of protein transduction, DNA transfection,
and structures between different organisms**

|  | Mammalian cells | Paramecium | Rotifer |
|---|---|---|---|
| Protein transduction | + | + | + |
| DNA transfection | + | + | + |
| Organism type | Single-celled | Single-celled | Multi-celled |
| Outer structure of cell membrane | − | + (pellicles) | + (cuticle) |

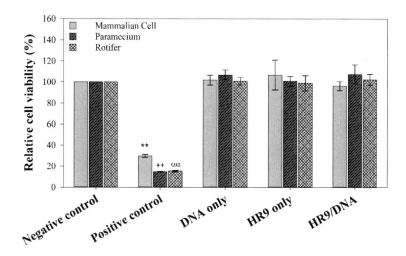

Figure 3. Cell viability of the HR9/DNA treatment. A549 cells, paramecia, and rotifers were treated with DNA only, HR9 only, and HR9/DNA complexes. Three types of specimens without any treatments served as a negative control. A549 cells and rotifers were treated with DMSO as a positive control, while paramecia were treated with 70% alcohol (EtOH) as a positive control. The MTT assay was used to evaluate cytotoxicity after these treatments. Each group was compared with the negative control. Significant differences were determined at P < 0.01 (**, ††, and αα) between the negative control and various treatments in each organisms.

To evaluate any cytotoxicity caused by HR9 or HR9/DNA complexes, A549 cells, paramecia, and rotifers were treated with the plasmid DNA only, HR9 only, and HR9/DNA complexes and subjected to the MTT assay (Figure 3). Organisms treated with 100% DMSO in A549 and rotifers or 70% alcohol in paramecia as the positive control illustrated distinct reduction of viabilities compared to the negative control (Figure 3). However, viability in treatment groups was not different from those in negative controls (Figure 3). These results suggested that neither HR9 nor HR9/DNA complexes cause cytotoxicity in these organisms.

## DISCUSSION

In this study, we demonstrated that HR9 could not only enter mammalian cells, paramecia, and rotifers by itself, but also deliver plasmid DNAs as cargoes into these different organisms. HR9-delivered DNAs still contained their bioactivities after cellular internalization. Potent transfection efficiencies, significant gene expression intensities, and harmless transfection processes revealed that CPPs (HR9) would be a good tool for transgenic protists studies.

Macromolecules, such as DNAs, RNAs, and proteins, were prevented to get into cells by plasma membranes because of the hydrophobic bilayer structures forming the perfect barrier. Paramecium, a ciliated unicellular organism, contains the extracellular structure called pellicles surrounding the plasma membrane (Table 1) [36-38]. This cortical ultrastructure made by elastic proteins forms three 2-dimentionally arrayed system and provides extra protections to stabilize cellular form and integrity [37, 38]. Similar to paramecium, rotifers possess outer structures of plasma membrane named cuticle (Table 1) [39-42]. Cuticles which are composed of sclerotized proteins to form multiple plates combine with a syncytial hypodermis to form the major integuments of rotifers. Maintaining shapes and osmosis, excluding external molecules and danger, providing basal laminate to spines, ridges, or other ornamentation, and offering a morphological diagnosis and identification among species are the roles and functions that cuticles play [41, 42]. Theoretically, it should be harder for exogenous molecules to penetrate plasma membrane in paramecia and rotifers. However, we indicated that CPPs displayed a complete departure from these limits and delivered macromolecules into cells/organisms (Figure 1 and 2; Table 1) [12, 15, 16, 23-28]. Pellicles, cuticle, and even multi-celled structures are not the keys for blocking CPPs' penetration.

Transfected genes in cells are easily degraded and lose their bioactivities via endocytosis-mediated internalization. Exogenous nucleic acids will be degraded by *lysosomal acidification* mechanisms unless they are able to escape from lysosomes or bypass endocytosis pathway while entering [43]. Likewise, paramecia and rotifers, single-celled and multi-celled *aquatic protozoa*, respectively, utilize similar intake routes and digest exogenous molecules by their vacuoles and stomachs [44, 45]. However, HR9 entered cells by the direct translocation mechanism and the cargoes that HR9 took and HR9 itself would not be trapped in lysosomes [14, 15]. Besides, HR9/cargo complexes would enter and stay in nuclei at the end of intracellular trafficking [46]. This is the good news for DNA transfection because exogenous genes could be transported into the inherited center directly and decrease risks of functional loss. CPP/DNA complexes internalized by cells via endocytosis showed lower transfection efficiencies, which supports this hypothesis [14-16, 24, 25, 46]. Here, we exhibited that cationic CPPs (HR9)-mediated gene delivery was able to increase transfection efficiencies, upregulate the gene expression, and decrease the injuries and immune responses [47] of transgenic targets with only one and simple transfection protocol in all tested organisms (mammalian cells, paramecia, and rotifers). Therefore, CPPs were potent tools for transfection and could be widely applied in the future.

# ACKNOWLEDGMENTS

We thank Ms Pin-Chine Liu, the Supply Center of Living Materials for High School Biology Experiment, Chenggong High School, Taipei, Taiwan, for provision of paramecia and Dr. Ilya N. Skovorodkin for the pGFP-actin1-1 plasmid. We also thank Daniel Goldman (University of Michigan, Ann Arbor, MI, USA) for provision of the pCS2+ DsRed plasmid. This work was supported by Grant Number Most106-2320-B-320-001-MY3 (to B.R.L.) from the Ministry of Science and Technology of Taiwan.

# REFERENCES

[1]   Stepanenko AA and Heng HH (2017) Transient and stable vector transfection: Pitfalls, off-target effects, artifacts. *Mutat Res/Reviews in Mutation Research* 773:91-103. doi: https://doi.org/10.1016/j.mrrev.2017.05.002.

[2]   Argani H (2019) Genome engineering for stem cell transplantation. *Exp Clin Transplant* 17:31-37. doi: 10.6002/ect.MESOT2018.L34.

[3]   Luo D and Saltzman WM (2000) Synthetic DNA delivery systems. *Nat Biotechnol* 18:33-7. doi: 10.1038/71889.

[4]   Haridhasapavalan KK, Borgohain MP, Dey C, Saha B, Narayan G, Kumar S and Thummer RP (2019) An insight into non-integrative gene delivery approaches to generate transgene-free induced pluripotent stem cells. *Gene* 686:146-159. doi: 10.1016/j.gene.2018.11.069.

[5]   Abdul-Cader MS, Amarasinghe A, Palomino-Tapia V, Ahmed-Hassan H, Bakhtawar K, Nagy E, Sharif S, Gomis S and Abdul-Careem MF (2018) In ovo CpG DNA delivery increases innate and adaptive immune cells in respiratory, gastrointestinal and immune systems post-hatch correlating with lower infectious laryngotracheitis virus infection. *PLoS One* 13:e0193964. doi: 10.1371/journal.pone.0193964.

[6]   D'Mello S, Salem AK, Hong L and Elangovan S (2016) Characterization and evaluation of the efficacy of cationic complex mediated plasmid DNA delivery in human embryonic palatal mesenchyme cells. *J Tissue Eng Regen Med* 10:927-937. doi: 10.1002/term.1873.

[7]   Subia B, Reinisalo M, Dey N, Tavakoli S, Subrizi A, Ganguli M and Ruponen M (2019) Nucleic acid delivery to differentiated retinal pigment epithelial cells using cell-penetrating peptide as a carrier. *Eur J of Pharm Biopharm* 140:91-99. doi: https://doi.org/10.1016/j.ejpb.2019.05.003.

[8] Brooks H, Lebleu B and Vivès E (2005) Tat peptide-mediated cellular delivery: back to basics. *Adv Drug Deliv Rev* 57:559-577. doi: https://doi.org/10.1016/j.addr.2004.12.001.

[9] Guidotti G, Brambilla L and Rossi D (2017) Cell-penetrating peptides: from basic research to clinics. *Trends Pharmacol Sci* 38:406-424. doi: https://doi.org/10.1016/j.tips.2017.01.003.

[10] Pooga M and Langel U (2015) Classes of cell-penetrating peptides. *Methods Mol Biol* 1324:3-28. doi: 10.1007/978-1-4939-2806-4_1.

[11] Hu JW, Liu BR, Wu CY, Lu SW and Lee HJ (2009) Protein transport in human cells mediated by covalently and noncovalently conjugated arginine-rich intracellular delivery peptides. *Peptides* 30:1669-78. doi: 10.1016/j.peptides.2009.06.006.

[12] Huang Y-W, Lee H-J, Tolliver LM and Aronstam RS (2015) Delivery of nucleic acids and nanomaterials by cell-penetrating peptides: opportunities and challenges. *Biomed Res Int* 2015:834079-834079. doi: 10.1155/2015/834079.

[13] Borrelli A, Tornesello AL, Tornesello ML and Buonaguro FM (2018) Cell penetrating peptides as molecular carriers for anti-cancer agents. *Molecules* (Basel, Switzerland) 23:295. doi: 10.3390/molecules23020295.

[14] Liu BR, Huang YW, Winiarz JG, Chiang HJ and Lee HJ (2011) Intracellular delivery of quantum dots mediated by a histidine- and arginine-rich HR9 cell-penetrating peptide through the direct membrane translocation mechanism. *Biomaterials* 32:3520-37. doi: 10.1016/j.biomaterials.2011.01.041.

[15] Liu BR, Lin M-D, Chiang H-J and Lee H-J (2012) Arginine-rich cell-penetrating peptides deliver gene into living human cells. *Gene* 505:37-45. doi: https://doi.org/10.1016/j.gene.2012.05.053.

[16] Liu BR, Liou JS, Chen YJ, Huang YW and Lee HJ (2013) Delivery of nucleic acids, proteins, and nanoparticles by arginine-rich cell-penetrating peptides in rotifers. *Mar Biotechnol* (NY) 15:584-95. doi: 10.1007/s10126-013-9509-0.

[17] Wadia JS and Dowdy SF (2002) Protein transduction technology. *Curr Opin Biotechnol* 13:52-6.

[18] Tunnemann G, Ter-Avetisyan G, Martin RM, Stockl M, Herrmann A and Cardoso MC (2008) Live-cell analysis of cell penetration ability and toxicity of oligo-arginines. *J Pept Sci* 14:469-76. doi: 10.1002/psc.968.

[19] Pipes BL, Vasanwala FH, Tsang TC, Zhang T, Luo P and Harris DT (2005) Brief heat shock increases stable integration of lipid-mediated DNA transfections. *Biotechniques* 38:48, 50, 52. doi: 10.2144/05381 bm05.

[20] Gomaa F, Garcia PA, Delaney J, Girguis PR, Buie CR and Edgcomb VP (2017) Toward establishing model organisms for marine protists: Successful transfection protocols for *Parabodo caudatus* (Kinetoplastida: Excavata). *Environ Microbiol* 19:3487-3499. doi: 10.1111/1462-2920.13830.

[21] Ohmachi M, Fujiwara Y, Muramatsu S, Yamada K, Iwata O, Suzuki K and Wang DO (2016) A modified single-cell electroporation method for molecule delivery into a motile protist, *Euglena gracilis*. *J Microbiol Methods* 130:106-111. doi: 10.1016/j.mimet. 2016.08.018.

[22] Li W, Ding H, Zhang X, Cao L, Li J, Gong P, Li H, Zhang G, Li S and Zhang X (2012) The viral RNA-based transfection of enhanced green fluorescent protein (EGFP) in the parasitic protozoan *Trichomonas vaginalis*. *Parasitol Res* 110:1305-10. doi: 10.1007/ s00436-011-2620-0.

[23] Chen CP, Chou JC, Liu BR, Chang M and Lee HJ (2007) Transfection and expression of plasmid DNA in plant cells by an arginine-rich intracellular delivery peptide without protoplast preparation. *FEBS Lett* 581:1891-7. doi: 10.1016/j.febslet. 2007.03.076.

[24] Chen YJ, Liu BR, Dai YH, Lee CY, Chan MH, Chen HH, Chiang HJ and Lee HJ (2012) A gene delivery system for insect cells mediated by arginine-rich cell-penetrating peptides. *Gene* 493:201-10. doi: 10.1016/j.gene.2011.11.060.

[25] Dai YH, Liu BR, Chiang HJ and Lee HJ (2011) Gene transport and expression by arginine-rich cell-penetrating peptides in Paramecium. *Gene* 489:89-97. doi: 10.1016/j.gene.2011.08.011.

[26] Hou YW, Chan MH, Hsu HR, Liu BR, Chen CP, Chen HH and Lee HJ (2007) Transdermal delivery of proteins mediated by non-covalently associated arginine-rich intracellular delivery peptides. *Exp Dermatol* 16:999-1006. doi: 10.1111/j.1600-0625.2007.00622.x.

[27] Liu BR, Chou JC and Lee HJ (2008) Cell membrane diversity in noncovalent protein transduction. *J Membr Biol* 222:1-15. doi: 10.1007/s00232-008-9096-6.

[28] Liu BR, Huang YW, Aronstam RS and Lee HJ (2015) Comparative mechanisms of protein transduction mediated by cell-penetrating peptides in prokaryotes. *J Membr Biol* 248:355-68. doi: 10.1007/s00232-015-9777-x.

[29] Liu BR, Huang YW, Aronstam RS and Lee HJ (2016) Identification of a short cell-penetrating peptide from bovine lactoferricin for intracellular delivery of DNA in human A549 cells. *PLoS One* 11:e0150439. doi: 10.1371/journal.pone.0150439.

[30] Liu BR, Huang YW and Lee HJ (2013) Mechanistic studies of intracellular delivery of proteins by cell-penetrating peptides in cyanobacteria. *BMC Microbiol* 13:57. doi: 10.1186/1471-2180-13-57.

[31] Lu SW, Hu JW, Liu BR, Lee CY, Li JF, Chou JC and Lee HJ (2010) Arginine-rich intracellular delivery peptides synchronously deliver covalently and noncovalently linked proteins into plant cells. *J Agric Food Chem* 58:2288-94. doi: 10.1021/jf903039j.

[32] Takenaka Y, Haga N, Harumoto T, Matsuura T and Mitsui Y (2002) Transformation of *Paramecium caudatum* with a novel expression vector harboring codon-optimized GFP gene. *Gene* 284:233-40.

[33] Strahl C and Blackburn EH (1994) The effects of nucleoside analogs on telomerase and telomeres in Tetrahymena. *Nucleic Acids Res* 22:893-900. doi: 10.1093/nar/22.6.893.

[34] Kim GY, Moon JM, Han JH, Kim KH and Rhim H (2011) The sCMV IE enhancer/promoter system for high-level expression and

efficient functional studies of target genes in mammalian cells and zebrafish. *Biotechnol Lett* 33:1319-26. doi: 10.1007/s10529-011-0589-5.

[35] Suhr ST, Ramachandran R, Fuller CL, Veldman MB, Byrd CA and Goldman D (2009) Highly-restricted, cell-specific expression of the simian CMV-IE promoter in transgenic zebrafish with age and after heat shock. *Gene Expr Patterns* 9:54-64. doi: 10.1016/j.gep.2008.07.002.

[36] Hufnagel LA (1969) Cortical ultrastructure of *Paramecium aurelia*. Studies on isolated pellicles. *J Cell Biol* 40:779-801. doi: 10.1083/jcb.40.3.779.

[37] Allen RD, Aihara MS and Fok AK (1998) The striated bands of paramecium are immunologically distinct from the centrin-specific infraciliary lattice and cytostomal cord. *J Eukaryot Microbiol* 45:202-9.

[38] Asterita H and Marsland D (1961) The pellicle as a factor in the stabilization of cellular form and integrity: effects of externally applied enzymes on the resistance of Blepharisma and Paramecium to pressure-induced cytolysis. *J Cell Comp Physiol* 58:49-61.

[39] Brodie AE (1970) Development of the cuticle in the rotifer *Asplanchna brightwelli*. *Z Zellforsch Mikrosk Anat* 105:515-25.

[40] Klusemann J, Kleinow W and Peters W (1990) The hard parts (trophi) of the rotifer mastax do contain chitin: evidence from studies on *Brachionus plicatilis*. *Histochemistry* 94:277-83.

[41] Schramm U (1978) Studies on the ultrastructure of the integument of the rotifer *Habrotrocha rosa* Donner (Aschelminthes). *Cell Tissue Res* 189:167-77.

[42] Zawierucha K, P GA, Buda J, Uetake J, Janko K and Fontaneto D (2018) Tardigrada and Rotifera from moss microhabitats on a disappearing Ugandan glacier, with the description of a new species of water bear. *Zootaxa* 4392:311-328. doi: 10.11646/zootaxa.4392.2.5.

[43] Underhill DM and Goodridge HS (2012) Information processing during phagocytosis. *Nat Rev Immunol* 12:492-502. doi: 10.1038/nri3244.

[44] Ramoino P (1992) Food ingestion and egestion in mating reactive populations of *Paramecium primaurelia. Boll Soc Ital Biol Sper* 68:535-41.

[45] Sawada M and Enesco HE (1984) A study of dietary restriction and lifespan in the rotifer *Asplanchna brightwelli* monitored by chronic neutral red exposure. *Exp Gerontol* 19:329-34.

[46] Liu BR, Lo SY, Liu CC, Chyan CL, Huang YW, Aronstam RS and Lee HJ (2013) Endocytic trafficking of nanoparticles delivered by cell-penetrating peptides comprised of nona-arginine and a penetration accelerating sequence. *PLoS One* 8:e67100. doi: 10.1371/journal.pone.0067100.

[47] Carter E, Lau CY, Tosh D, Ward SG, Mrsny RJ (2013) Cell penetrating peptides fail to induce an innate immune response in epithelial cells in vitro: implications for continued therapeutic use. *Eur J Pharm Biopharm* 85:12-9. doi: 10.1016/j.ejpb.2013.03.024.

# INDEX

## D

## E

## L

## M

**P**

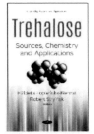